Computational Statistics and Applications

Edited by Ricardo López-Ruiz

Published in London, United Kingdom

IntechOpen

Supporting open minds since 2005

Computational Statistics and Applications
http://dx.doi.org/10.5772/intechopen.95652
Edited by Ricardo López-Ruiz

Contributors
Lalpawimawha, Arvind Pandey, Nilgün Yildiz, Mu Yue, D.K Ghosh, Yüksel Akay Ünvan, Ulviyya Nahmatli, Tomáš Mrkvička, Wenlin Dai, Stavros Athanasiadis, Adji Achmad Rinaldo Fernandes, Solimun, Nurjannah, Masahiro Kuroda, Amod Kumar, Catherine Chunling Liu, Junshan Shen, Chong Zhong, Zhihua Ma

First published in London, United Kingdom, 2022 by IntechOpen
IntechOpen is the global imprint of INTECHOPEN LIMITED, registered in England and Wales, registration number: 11086078, 5 Princes Gate Court, London, SW7 2QJ, United Kingdom
Printed in Croatia

British Library Cataloguing-in-Publication Data
A catalogue record for this book is available from the British Library

Additional hard and PDF copies can be obtained from orders@intechopen.com

Computational Statistics and Applications
Edited by Ricardo López-Ruiz
p. cm.
Print ISBN 978-1-83969-782-1
Online ISBN 978-1-83969-783-8
eBook (PDF) ISBN 978-1-83969-784-5

Meet the editor

Ricardo López-Ruiz, MS, Ph.D., is an associate professor in the Department of Computer Science and Systems Engineering, Faculty of Science, University of Zaragoza, Spain. He is also an associate researcher in Complex Systems at the School of Mathematics, University of Zaragoza. Previously, he worked as a lecturer at the University of Navarra, the Public University of Navarra, and UNED Calatayud, all in Spain. He completed his postdoc with Prof. Yves Pomeau at the École Normale Supérieure, Paris, France, and with Prof. Gabriel Mindlin at the University of Buenos Aires, Argentina. His areas of interest include statistical complexity and nonlinear models, chaotic maps and applications, multiagent systems, econophysics, big data, and artificial intelligence techniques.

Contents

Preface

Nature evolves mainly in a statistical way. Different strategies, formulas, and conformations are continuously confronted in the natural processes. Some of them are selected and then the evolution continues with a new loop of confrontation for the next generation of phenomena and living beings. Failings are corrected without a previous program or design. The new options generated by different statistical and random scenarios lead to solutions for surviving the present conditions. This is the general panorama for all scrutiny levels of the life cycles. Over three parts, this book examines different statistical questions and techniques in the context of machine learning and clustering methods, the frailty models used in survival analysis, and other studies of statistics applied to diverse problems.

The first section of the book presents different techniques and methods applied to clustering and machine learning. In Chapter 1, Dai et al. propose a framework for a clustering procedure based on functional rankings or depth. In Chapter 2, Fernandes et al. discuss the reasons to use dummy variables in cluster analysis. In Chapter 3, Yue is concerned with sparse boosting-based machine learning methods in different high-dimensional problems. Chapter 4 by Kuroda presents how to accelerate the convergence of the EM algorithm and apply it to mixture model estimation.

The second section of the book addresses the question of frailty models usually applied to survival analysis. In Chapter 5, Zhong et al. propose a generalized shared frailty model and develop a survival function to model the dependency among the baseline survival functions. Chapter 6 by Pandey and Lalpawimawha introduces a new frailty model with exponential power and generalized Rayleigh as baseline distributions.

The last section of the book presents the use of computational statistics in different contexts and problems. In Chapter 7, Yildiz depicts the use of the network meta-analysis tool through an example from diabetes. In Chapter 8, Ghosh constructs an N-ary variance balance design by using different techniques. In Chapter 9, Kumar proposes an improved randomized response model for the simultaneous estimation of population means of two quantitative sensitive variables. In Chapter 10, Ünvan and Nahmatli examine the causal relationship between imports, exports, and Exim bank loans in the Turkish economy.

As the editor of this book, I would like to thank all the contributing authors and reviewers. I am also grateful to the staff at IntechOpen, particularly Author Service Manager Ms. Romina Rovan. At this time when the omicron variant of the coronavirus continues to plague the world, I want to dedicate this book to all the teachers I had in the CP El Castelar of Villafranca and the IES Marqués de Villena of Marcilla. Finally, I acknowledge the support of my family, friends, and advisors.

Ricardo López-Ruiz
University of Zaragoza,
Spain

Section 1

Clustering and Machine Learning Methods

Chapter 1

A New Functional Clustering Method with Combined Dissimilarity Sources and Graphical Interpretation

Wenlin Dai, Stavros Athanasiadis and Tomáš Mrkvička

Abstract

Clustering is an essential task in functional data analysis. In this study, we propose a framework for a clustering procedure based on functional rankings or depth. Our methods naturally combine various types of between-cluster variation equally, which caters to various discriminative sources of functional data; for example, they combine raw data with transformed data or various components of multivariate functional data with their covariance. Our methods also enhance the clustering results with a visualization tool that allows intrinsic graphical interpretation. Finally, our methods are model-free and nonparametric and hence are robust to heavy-tailed distribution or potential outliers. The implementation and performance of the proposed methods are illustrated with a simulation study and applied to three real-world applications.

Keywords: depth, insurance, intrinsic graphical interpretation, robustness, statistical rankings

1. Introduction

Cluster analysis is a critical step in exploratory data analysis intended to identify homogeneous subgroups among observations. Cluster analysis is also widely used for functional data in tasks such as the classification of electrocardiogram curves in the diagnosis of cardiovascular ischemic diseases [1] and the extraction of representative wind behavior [2, 3]. The various functional data clustering methods described in the literature can generally be categorized into two subgroups: distance-based methods and filtering-based methods.

The distance-based methods involve the construction of a distance matrix with a specific metric; the clustering results may be derived with hierarchical or centroid-based clustering tools [3, 4]. The filtering-based methods involve the approximation of the curves with linear combinations of finite basis functions, such as splines and functional principal components, and the cluster analysis is conducted based on the coefficients or scores of finite dimensions [5–7]. The focus of this study is on distance-based methods. In this paper, we propose a new family of clustering algorithms based on the chosen functional ordering. The dissimilarity matrix is constructed via the chosen functional ordering, which is applied to the set of differences of all pairs of the functional data under investigation. Various functional ordering can be chosen, but we concentrate on orderings with intrinsic graphical interpretation. But any ordering that treats the sources equally can be used, including the modified band depth [8] and the simplicial band depth [9]. The choice of functional ordering with intrinsic graphical interpretation allows us to show the resulting clusters and a central region that attains a natural interpretation. I.e., All functions contained in the central region do not leave the plot of the central region, and all functions not contained in the central region leave the plot of the central region in at least one point. It has to be mentioned that the classical functional orderings mentioned above do not satisfy this natural condition, and therefore we will concentrate on functional orderings defined in [10].

Functional data differ in various ways, such as in magnitude, shape, phase, and dependence structure, and hence they are difficult to analyze when clusters exist from multiple perspectives. The existing methods either focus on a single type of variation or pool the various sources of variation with weightings that rely on a delicate selection procedure. Without balancing, the clustering results could be dominated by the component with the greatest absolute variation. In order to achieve some balancing between the various sources, it is possible to standardize the curves before applying existing methods, such as k-means or model-based methods. By "standardization", we mean that the marginal empirical distributions are standardized so that they have zero mean and unit variance. This approach is used in the simulation study in order to compare the performance of existing methods with the proposed methods.

Since the proposed procedure applies functional ordering, such that every part of the function is treated equally, the different sources of variation are combined in an equal manner. For univariate cases, it may combine the raw curves and the derivatives equally to measure the magnitude and shape variation simultaneously. For multivariate cases, it may combine the marginal curves and the covariance functions equally to account for both marginal and joint variation among curves. Furthermore, the proposed method provides a reasonable graphical interpretation of the clustering result. Finally, it inherits the robustness of functional orderings and can stably recover the clusters when abnormal observations contaminate the data.

The remainder of this paper is organized as follows. In Section 2, we define the new proposed procedure with an arbitrary functional ordering. Further, we review several functional orderings already defined in [10] which satisfy the intrinsic graphical interpretation. Finally, we study the metric properties of derived dissimilarity. In Section 3, we describe the simulation studies we conducted to assess the performance of the proposed methods and compare them with some existing competitors in cases where the combination of the various sources is of interest. In Sections 4–6, we demonstrate the effectiveness of our method with three real-world examples. The proposed methods will be available soon in the R package GET.

2. Description of methods

2.1 Dissimilarity matrix

Assume that the functions $f_i(x), i = 1, \dots, s$ are observed at a fixed set of points $x_1, \dots, x_d$, so that the functions can be represented as d-dimensional vectors $\mathbf{T}_i, i = 1, \dots, s$. If the functions of interest are not observed at the same set of points, a nonparametric smoothing method can be applied to address the situation.

To induce dissimilarity measure from functional ordering, we construct the set of functional differences:

$$D_f = \{df_{ii'} = f_i(x) - f_i'(x), \quad i, i' = 1, \dots, s\}.$$

We remark here that $df \equiv 0$ is an element of D_f. We then apply a functional ordering to D_f and obtain the induced measure of centrality of $df_{ii'} = f_i(x) - f_i'(x)$ as $M_{ii'}$. Finally, the dissimilarity between $f_i(x)$ and $f_{i'}(x)$ is defined as $d_{ii'} = 1 - M_{ii'}$, and this forms the dissimilarity matrix of $\{f_i\}_{i=1}^{s}$. Such an ordering can take the form of any functional depth notions or rankings in the literature, such as the band depth and modified band depth [8], the simplicial band depth [9], the spatial functional depth [11], or the curve depth [12]. These notions naturally give equal treatment to the variations at each design point, compared with the norm-based methods such as L_1 or L_2 distances.

After a dissimilarity matrix is established, the partitioning around medoids procedure can be used to determine the given number of clusters. This produces a family of clustering algorithms that depends on the choice of the functional ordering.

In the following, we will discuss the possible choices of functional ordering. First, we assume functional orderings, which take different sources of the data variability equally. We call such ordering combined functional ordering. Such an approach is useful when the investigator wants to join different information about the data and combine them in one universal procedure. Second, we review several functional orderings which satisfy the intrinsic graphical interpretation.

Our proposed procedure then consists of the following steps:

1. Choose the appropriate data sources (e.g., raw data, derivative and second derivative)

2. Choose the functional ordering, which allows for intrinsic graphical interpretation and which gives the same weight to every chosen source (e.g., the studentized maximum ordering, the area rank ordering).

3. Compute the dissimilarity matrix

4. Apply partitioning around medoids

5. Plot the resulted clusters together with their central region with intrinsic graphical interpretation.

2.2 Combined functional ordering

We consider now functions $T_i(x), i = 1, \dots, s'$ and specify their combined functional ordering. Various perspectives, such as different magnitudes and different shapes of the functions, can be used to order the functions. Here we provide a general method to combine these different perspectives in an equal manner. As suggested by [13], data transformation is an effective method to convert different types of variation into types that are easy to handle by the functional depth. Hence, various transformations could be applied to the raw functions to obtain the transformed data sets of interest, such as $V_1, \dots, V_k$. These transformations are computed in the same fixed set of points $x_1, \dots, x_d$; for instance, shifting each curve to zero means eliminates the magnitude variation, normalizing the centered curves by their L_2 norms, respectively, to extract pure shape information. In the case of multivariate functional data, each component of the data and their transformation could be treated similarly. Also, the covariance function between the components can be added to take into account the dependence structure.

We denote with $V_k(T_{ij})$ the resultant curves of T_{ij} via the transformation V_k, and we can express the long vector as:

$$\mathbf{T}_i = (V_1(T_{i1}), \ldots, V_1(T_{id}), \ldots, V_k(T_{i1}), \ldots, V_k(T_{id})), \quad i = 1, \ldots, s' \tag{1}$$

We can then apply to them the corresponding ordering and hence construct the dissimilarity matrix. Note that each of the orderings to be introduced considers each element equally by ranking or scaling, so the desired perspectives of ordering are considered and treated equally in such a combined ordering. To enhance the interpretability of the clustering results, we focus only on the notions that satisfy the intrinsic graphical interpretation.

2.3 Functional ordering with intrinsic graphical interpretation

The following definition specifies the properties of *the global envelope that has an intrinsic graphical interpretation with respect to an ordering*. This definition was already used in [10] to define global envelope tests and central regions with graphical interpretation.

Definition 1: Assume a general ordering $\prec$ of the vectors $\mathbf{T}_i, i = 1, \ldots, s'$, that is induced by a univariate measure M_i. That is, $M_i \geq M_j$ iff $\mathbf{T}_i \prec \mathbf{T}_j$, which means that $\mathbf{T}_i$ is less extreme or as extreme as $\mathbf{T}_j$. (The smaller the measure M_i, the more extreme $\mathbf{T}_i$.) The $100(1-\alpha)\%$ global envelope $\left[\mathbf{T}_{\text{low}j}^{(\alpha)}, \mathbf{T}_{\text{upp}j}^{(\alpha)}\right]$ has *intrinsic graphical interpretation* (IGI) with respect to the ordering $\prec$ if:

1. $m_{(\alpha)} \in \mathbb{R}$ is the largest of the M_i such that the number of those i for which $M_i < m_{(\alpha)}$ is less than or equal to $\alpha s'$;

2. $T_{ij} < \mathbf{T}_{\text{low}j}^{(\alpha)}$ or $T_{ij} > \mathbf{T}_{\text{upp}j}^{(\alpha)}$ for some $j = 1, \ldots, d$ iff $M_i < m_{(\alpha)}$ for every $i = 1, \ldots, s'$;

3. $\mathbf{T}_{\text{low}j}^{(\alpha)} \leq T_{ij} \leq \mathbf{T}_{\text{upp}j}^{(\alpha)}$ for all $j = 1, \ldots, d$ iff $M_i \geq m_{(\alpha)}$ for every $i = 1, \ldots, s'$.

Let us call *the ordering with intrinsic graphical interpretation* such ordering, for which exists a global envelope with IGI with respect to this ordering. Remark here that $m_{(\alpha)}$ is not exactly the α quantile of M_i and that points 2 and 3 are equivalent. We kept points 2 and 3 to show the interpretability of the IGI. The simple ordering criterion based on L_∞ distance, $M_i = \max_j |T_{ij} - \overline{T}_{\cdot j}|$, clearly satisfies such a property, but it does not account for the changes in the marginal distribution of $T_{\cdot j}$ for different values of j [14, 15]. To address this problem, Myllymäki et al. [14] proposed studentized and directional quantile scaling of the maximum ordering, which also satisfies IGI. Furthermore, [15, 16] simultaneously defined extreme rank length ordering, which is based on the number of the most extreme pointwise ranks and satisfies IGI. Finally, [10] extended this family with continuous rank ordering, which is based on the continuous extension of pointwise ranking, and area rank ordering, which is based on the area with the most extreme continuous ranks. To the best of our knowledge, no other functional (respective multivariate) orderings satisfy IGI.

The definitions of all previously mentioned orderings are given in [10]. For the sake of completeness, we provide here a short list of these definitions.

2.3.1 Extreme rank length ordering

Let $r_{1j}, r_{2j}, \dots, r_{s'j}$ be the raw ranks of $T_{1j}, T_{2j}, \dots, T_{s'j}$, such that the smallest T_{ij} has rank 1. In the case of ties, the raw ranks are averaged. The two-sided pointwise ranks are then calculated as $R_{ij} = \min\left(r_{ij}, s' + 1 - r_{ij}\right)$. Consider now the vectors of pointwise ordered ranks $\mathbf{R}_i = \left(R_{i[1]}, R_{i[2]}, \dots, R_{i[d]}\right)$, where $\left\{R_{i[1]}, \dots, R_{i[d]}\right\} = \left\{R_{i1}, \dots, R_{id}\right\}$ and $R_{i[k]} \leq R_{i[k']}$ whenever $k \leq k'$. The extreme rank length measure of the vectors $\mathbf{R}_i$ is equal to:

$$E_i = \frac{1}{s'} \sum_{i'=1}^{s'} (\mathbf{R}_{i'} \prec \mathbf{R}_i) \tag{2}$$

where

$$\mathbf{R}_i' \prec \mathbf{R}_i \Leftrightarrow \exists n \leq d : R_{i'[k]} = R_{i[k]} \forall k < n, \quad R_{i'[n]} < R_{i[n]}.$$

The division by s' leads to normalized ranks that obtain values between 0 and 1. Consequently, the ERL measure corresponds to the extremal depth as defined in [16].

Let e_α be defined according to point 1 of Definition 2.3, and let $I_\alpha = \left\{i \in 1, \dots, s' : E_i \geq e_{(\alpha)}\right\}$ be the index set of vectors less extreme than or as extreme as e_α. Then, the $100(1-\alpha)\%$ global extreme rank length envelope (or global extreme rank length central region) induced by E_i is:

$$\mathbf{T}^{(\alpha)}_{\text{low}k} = \min_{i \in I_\alpha} T_{ik} \quad \text{and} \quad \mathbf{T}^{(\alpha)}_{\text{upp}k} = \max_{i \in I_\alpha} T_{ik} \quad \text{for} k = 1, \dots, d. \tag{3}$$

2.3.2 Global continuous rank ordering

The continuous rank measure is:

$$C_i = \min_{j=1,\dots d} c_{ij} / \lceil s'/2 \rceil,$$

where c_{ij} are the pointwise continuous ranks defined as:

$$c_{ij} = \sum_{i'} \mathbf{1}\left(T_{i'j} > T_{ij}\right) + \frac{T_{[i+1]j} - T_{ij}}{T_{[i+1]j} - T_{[i-1]j}} \qquad \text{for } i : T_{ij} \neq \max_{i'} T_{i'j}$$
$$\text{and } T_{ij} > \text{median}\left(T_{ij}\right),$$

$$c_{ij} = \exp\left(-\frac{T_{ij} - T_{[i-1]j}}{T_{[i-1]j} - \min_i T_{ij}}\right) \qquad \text{for } i : T_{ij} = \max_{i'} T_{i'j},$$

$$c_{ij} = \sum_{i'} \mathbf{1}\left(T_{i'j} < T_{ij}\right) + \frac{T_{ij} - T_{[i-1]j}}{T_{[i+1]j} - T_{[i-1]j}} \qquad \text{for } i : T_{ij} \neq \min_{i'} T_{i'j}$$
$$\text{and } T_{ij} < \text{median}\left(T_{ij}\right),$$

$$c_{ij} = \exp\left(-\frac{T_{[i+1]j} - T_{ij}}{\max_i T_{ij} - T_{[i+1]j}}\right) \qquad \text{for } i : T_{ij} = \min_{i'} T_{i'j}.$$

$$c_{ij} = R_{ij} \qquad \text{for } T_{ij} = \text{median}\left(T_{ij}\right),$$

Here, $T_{[i-1]j}$ and $T_{[i+1]j}$ denote the values of the functions, which are in a j-th element below and above T_{ij}, respectively (i.e., $T_{[i-1]j} = \max_{i':T_{i'j}<T_{ij}} T_{i'j}$ and $T_{[i+1]j} = \min_{i':T_{i'j}>T_{ij}} T_{i'j}$).

The $100(1-\alpha)\%$ global continuous rank envelope induced by C_i is constructed in the same manner as the global extreme rank length envelope.

2.3.3 Global area rank ordering

The area rank measure:

$$A_i = \frac{1}{\lceil s'/2 \rceil d} \sum_j \min\left(R_i, c_{ij}\right),$$

where.

$R_i = \min_j\{R_{ij}\}$ and R_{ij} are two-sided pointwise ranks defined above. The $100(1-\alpha)\%$ global area rank envelope induced by A_i is constructed in a manner similar to that of the global extreme rank length envelope.

2.3.4 Studentized maximum ordering

Because we construct a symmetric set of functions to compute the dissimilarity matrix, here we use only the symmetric studentized ordering. The above orderings are based on the whole distributions of $T_{\cdot j}, j = 1, \dots, d$. It is also possible to approximate the distribution from a few sample characteristics. The studentized maximum ordering approximates the distribution of $T_{\cdot j}, j = 1, \dots, d$ by the sample mean T_{0j} and sample standard deviation $\mathrm{sd}(T_{\cdot j})$. The studentized measure is:

$$S_i = \max_j \left| \frac{T_{ij} - T_{0j}}{\mathrm{sd}(T_{\cdot j})} \right|. \quad (4)$$

The $100(1-\alpha)\%$ global studentized envelope induced by S_i is defined by:

$$\mathbf{T}^{(l)}_{\mathrm{low}j} = T_{0j} - s_\alpha \mathrm{sd}(T_{\cdot j}) \quad \text{and} \quad \mathbf{T}^{(l)}_{\mathrm{upp}j} = T_{0j} + s_\alpha \mathrm{sd}(T_{\cdot j}) \quad \text{for} j = 1, \dots, d, \quad (5)$$

where s_α is taken according to point 1 of IGI.

2.4 Dissimilarity matrix based on the combined ordering

In this section, we validate the dissimilarity matrix construction defined in Section 2.1 for studentized measure by showing that $d_{ii'} = S_{ii'}$ is a metric and for global area rank measure by showing that $d_{ii'} = 1 - A_{ii'}$ is a semi-metric. The latter means that the $d_{ii'} = 1 - A_{ii'}$ satisfies all properties of metric, except for the triangular inequality, which is violated in specific cases. The metric properties are usually required when choosing the distance measure, but it is not necessary for the partitioning around medoids algorithm, which is used to calculate the clusters afterward. Furthermore, our simulation study demonstrates that these specific cases, where the triangular inequality of global area rank measure is not satisfied, are not realized by functions appearing in real data studies. Furthermore, we provide a thorough check of satisfaction of the triangular inequality for global area rank measure in our implementation of the algorithm. Thus in practice, a user can check this feature of the metric for particular data of interest. For any dataset considered by us in simulation and data studies, the triangular inequality was satisfied.

Theorem 1.1: Define the distance between $\mathbf{T}_i$ and $\mathbf{T}_{i'}$ as:

$$d_{ii'} = 1 - A_{ii'},$$

where $A_{ii'}$ is the global area rank measure of $T_i - T_{i'}$ on D_f. Then $d_{ii'}$ satisfies for any i, i':

1. Non-negativity: $d_{ii'} \geq 0$;

2. Identity of indiscernibles: $d_{ii'} = 0$ iff $\mathbf{T}_i = \mathbf{T}_{i'}$;

3. Symmetry: $d_{ii'} = d_{i'i}$.

Proof:

Non-negativity: *For the set D_f, there are s' curves. The set D_f contains a zero element, which is the deepest point of D_f. I.e. 0 is median in every coordinate. For the area ordering of these curves, we have that two-sided pointwise ranks of curve $\mathbf{T}_i - \mathbf{T}_{i'}$ is $R_{ii'j} \leq \lceil s'/2 \rceil$ and $R_{ii'} = \min_j \{R_{ii'1}, \dots, R_{ii'd}\} \leq \lceil s'/2 \rceil$. Hence, we have $A_{ii'} \leq 1$, i.e., $d_{ii'} \geq 0$.*

Identity of indiscernibles: *$d_{ii'} = 0 \Leftrightarrow A_{ii'} = 1 \Leftrightarrow R_{ii'j} = \lceil s'/2 \rceil$ for every $j = 1, \dots, d$ $\Leftrightarrow \mathbf{T}_i - \mathbf{T}_{i'}$ is the deepest curve of D_f $\Leftrightarrow \mathbf{T}_i = \mathbf{T}_{i'}$.*

Symmetry: *This property holds implicitely due to the symmetry of D_f.*

The fourth property of the metric, i.e.

4. Triangle inequality: $d_{ii'} + d_{i'k} \geq d_{ik}$, for any i, i' and k,

   is not satisfied when $\mathbf{T_i} \equiv t_i$ for every i. The results of our simulation study suggest that if the system of data provides enough crossings of functions, then the triangle inequality is satisfied.

Theorem 1.2: Define the distance between $\mathbf{T}_i$ and $\mathbf{T}_{i'}$ as:

$$d_{ii'} = S_{ii'},$$

where $S_{ii'}$ is the studentized measure of $T_i - T_{i'}$ on D_f. Then $d_{ii'}$ is a valid metric.

Proof:

The first three properties obviously hold for the studentized difference distance. We prove the triangle inequality for $d_{ii'}$. Note that $df \equiv 0$ is an element of D_f, and hence the sample mean $T_{0j} = 0$ for $j = 1, \dots, d$. Lets denote the sample standard deviation of the j-th element of D_f by $\mathrm{sd}(D_{\cdot j})$. Then, we have:

$$\begin{aligned} d_{ik} &= \max_j \left| \frac{T_{ij} - T_{kj} - 0}{\mathrm{sd}(D_{\cdot j})} \right| \leq \max_j \left\{ \left| \frac{T_{ij} - T_{i'j}}{\mathrm{sd}(D_{\cdot j})} \right| + \left| \frac{T_{i'j} - T_{kj}}{\mathrm{sd}(D_{\cdot j})} \right| \right\} \\ &\leq \max_j \left| \frac{T_{ij} - T_{i'j}}{\mathrm{sd}(D_{\cdot j})} \right| + \max_j \left| \frac{T_{i'j} - T_{kj}}{\mathrm{sd}(D_{\cdot j})} \right| \\ &= d_{ii'} + d_{i'k}. \end{aligned}$$

This completes the proof.

3. Simulation study

This section describes the intensive simulation studies we conducted to assess the empirical performance of the proposed clustering methods and compares this performance with those of the existing methods when the clusters demonstrate differences from various perspectives. For comparison, we also consider two clustering methods for functional data: the k-means methods available in the R package *fda.usc* [17] and the model-based clustering methods proposed by [18], which are available in the R package *fdapace* [19]. For the fairness of comparison, the standardization procedure is applied to normalize the empirical marginal distributions as described in Section 1 so that they can be combined equally.

Specifically, we consider the following five models on $t \in [0, 1]$:

- Class 1: $X(T) = 2T + e(T)$;
- Class 2: $X(T) = 2 - 2T + e(T)$;
- Class 3: $X(T) = 2\,\mathbf{1}(T > U) + e(T)$;
- Class 4: $X(T) = 1.5 + 2\,\mathbf{1}(T > U) + e(T)$;
- Class 5: $X(T) = 3 - 2.5T + e(T)$.

Here, U follows a uniform distribution on $[0.5, 0.6]$, and $e(T)$ is generated from a Gaussian process with zero mean and covariance function $\gamma(s,t) = \sigma^2 \exp\{-\phi|t-s|^{\nu}\}$, where $\sigma^2 = 0.2$, $\phi = 2$ and $\nu = 1$.

In addition, to assess the robustness of the proposed methods, we also consider another situation by replacing $e(T)$ with a multivariate-t distribution with two degrees of freedom, $t_2(\mu, \Sigma)$, where $\mu = 0$, and Σ is generated with $\gamma(s,t)$. The heavy

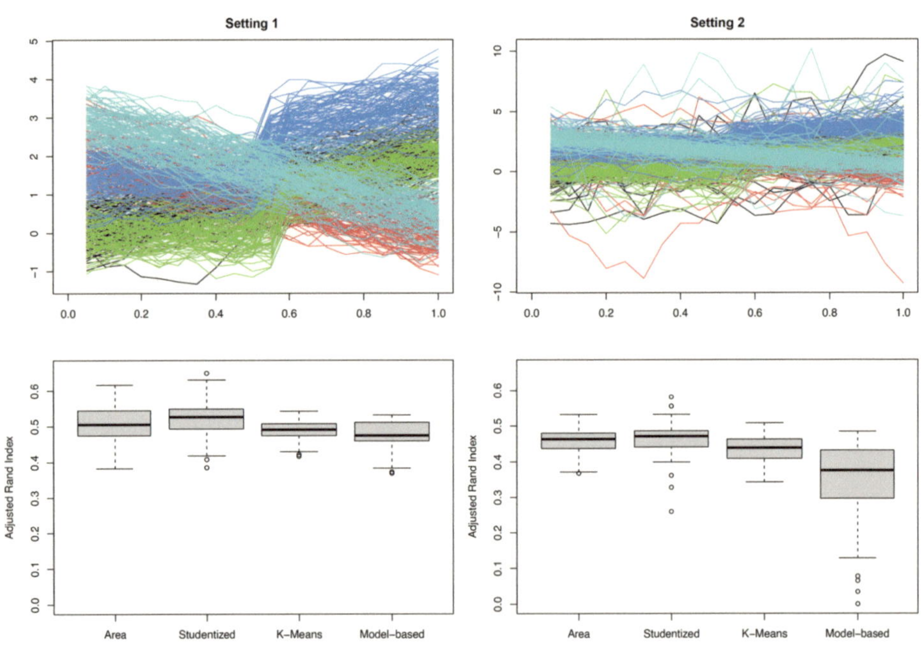

Figure 1.
Top panel: Realizations of two settings. Bottom panel: Adjusted Rand index of four clustering methods with the two settings.

DOI: http://dx.doi.org/10.5772/intechopen.100124

tail property of the marginal distribution allows the data to be viewed as contaminated by some outliers, which are commonly encountered in practice. We generate 100 samples for each of the five classes with 20 equally spaced design points; as a result, 500 curves are clustered into five groups. The top panel of **Figure 1** demonstrates one realization of the simulated samples under two settings. To account for both the magnitude and the shape variation among clusters, we make two

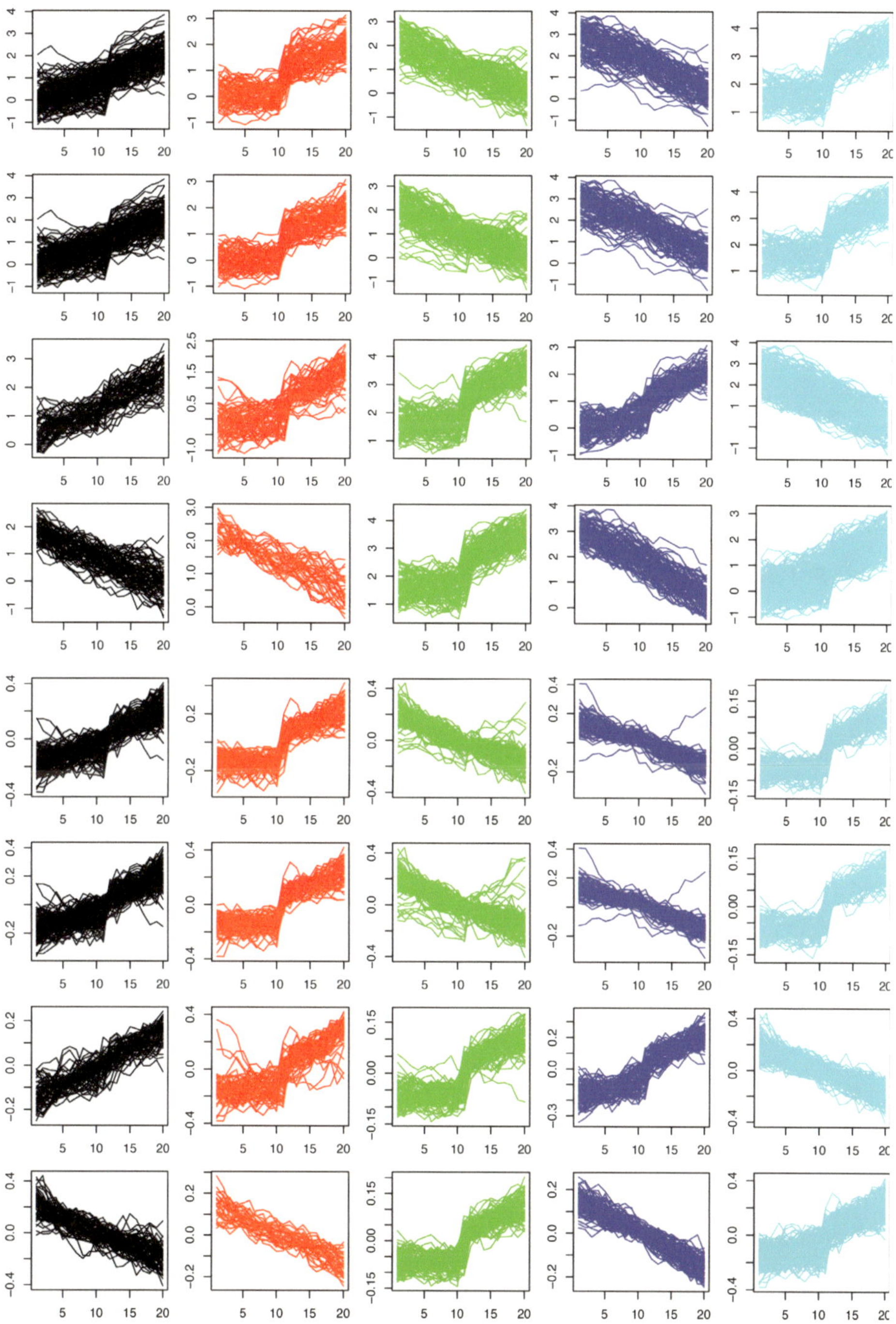

Figure 2.
Clusters for setting 1 visualized on raw curves (top panel) and normalized curves (bottom panel). In each panel, from top to bottom: Area, studentized, k-means, and model based.

transformations suggested by [13] to the raw curves, shifting the curves so that each has a zero mean and then normalizing the centered curves by their L_2 norm. We then bind the three components together as long vectors for clustering. For each run, we use the true number of clusters for all four methods and calculate the adjusted Rand index [20] to compare their clustering results. We repeat the procedure 100 times, and the results are reported in the bottom panel of **Figure 1**.

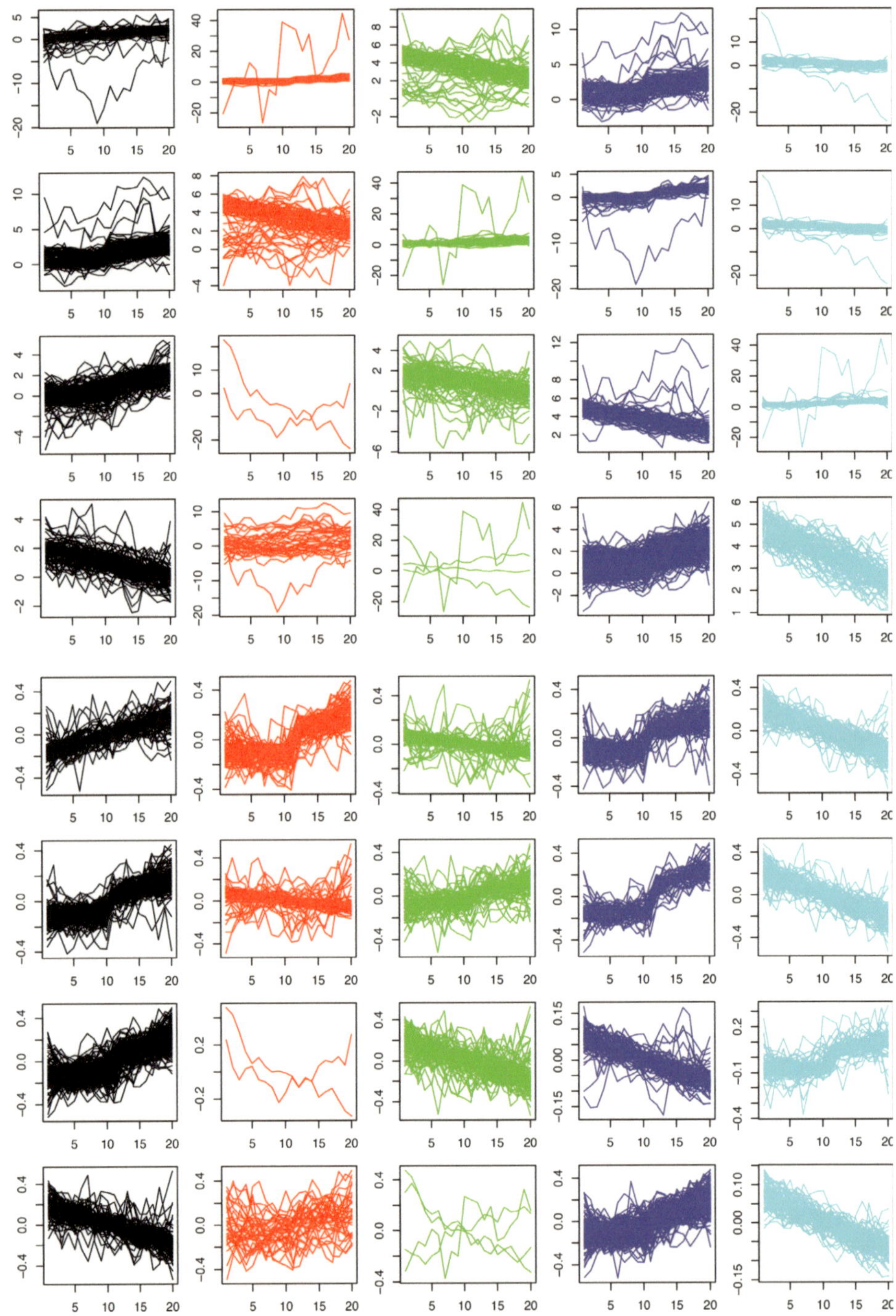

Figure 3.
Clusters setting 2 visualized on raw curves (top panel) and normalized curves (bottom panel). In each panel, from top to bottom: Area, studentized, k-means, and model based.

Note that in all cases of the simulation study, the triangular inequality of the area measure was satisfied for all combinations of curves.

Under the first setting, data are generated from a Gaussian process. With regard to the adjusted Rand index, the four methods are quite comparable but the proposed methods are slightly better than the other two. However, our methods recover much better the characteristics of the true clusters; see **Figure 2**, which illustrates one clustering result for each of the four methods with both raw curves and normalized curves. In contrast, the k-means method merges classes 2 and 5, and the model-based method merges classes 1 and 3.

As for the second setting, in which the marginal distribution becomes heavy-tailed, our methods obtain more robust clustering results than the other two methods and reach higher adjusted Rand indexes (**Figure 3**). The model-based method relies heavily on the Gaussian assumption and thus shows less satisfactory behavior. Again, our methods still accurately recover the patterns of each cluster, whereas the other two methods completely fail to reveal reasonable group structures. Specifically, both k-means and the model-based methods suggest a cluster with only a few curves, which indicates a clear misinterpretation of the situation.

4. Clustering of insurance penetration

Insurance consumption indicates the equilibrium of supply and demand of insurance products. For a given insurance market, the collection of total (Life and non-Life) yearly insurance consumption observations helps to explain the variation of insurance market development over time. A common measure of insurance consumption, and hence of insurance development, is insurance penetration (IP), defined as the ratio of insurance premiums on GDP. The pattern of the development variation is evident when one views the IP as a function of time, known as the IP curve.

In their effort to promote the European single insurance market through the integration process, European policymakers put emphasis on homogeneity and convergence aspects of development patterns of European insurance markets. That is equivalent to saying that they are interested in identifying a single group (cluster) of countries whose IP curves exhibit similarity in magnitude and shape. The clustering of European countries in terms of their IP curves provides a method for testing the magnitude and shape similarity of the insurance industry in Europe. In particular, functional clustering methods are appropriate for our data, given the time dependency in the observations.

IP curves (time-series data on IP) originated from the Swiss Re (2016) Database were analyzed by the proposed functional clustering (FC) method based on Area measure. The exploration concentrated on the IP curves of 34 European countries (EU and non-EU members) observed over 13 years between 2004 and 2016, that is, before, during, and post-financial and sovereign debt crises.

The FC method extracts the partitioning information from both the magnitude and the shape of IP curves. While the magnitude is captured in the IP curves, the shape is not straightforward to be detected. To this end, we performed two types of transformations on the raw IP curves to reveal their shape. First, the raw IP curves were centred relative to each country's average IP rate to mitigate the widely different magnitudes in the IP data. After this, the resulting centred IP curves were then normalized with their L2 norms to a unit norm (to have a length of 1). These transformations are proposed to extract shape information by [13] By normalizing the centred IP curves in this manner, we eliminate their amplitude signal, while we are only left with the shape signal of the raw IP curves.

For the FC method to run properly, the most suitable number of clusters must be determined. We chose 6 clusters even if the median value of all methods presented in the NBclust library of the R software is 5. Our choice is justified as it better serves the analysis and the characterization of the produced clusters.

Given the IP curves of each cluster, the FC method also provides a graphical representation, through the central regions, of the deepest central IP curves within each cluster. We are interested in the so-called marginal plot style approach of the clustering solution. This means that the central regions are computed separately for magnitude and shape to better express each cluster component's shape. Remark here that the proposed method also allows showing the central region with respect to the combined ordering with respect to the magnitude and shape together. The appearance of clusters is demonstrated by the deepest IP curve (solid curve) that corresponds to the medoid IP curve and the envelope of 50% central IP curves (gray area) that reflects the band where 50% of the IP curves surrounding the deepest are varied. See **Figures 4** and **5**. Note that the fraction of combinations of countries satisfying the triangular inequality with Area measure was 1 with respect to all combinations. With this visualization, we can describe the clusters that are produced by the FC method as follows:

Cluster 1: Developed insurance markets with middle-to-high IP levels and decreasing IP patterns in the whole period. This cluster includes Belgium, France, Ireland, Austria*, the UK, Portugal, Switzerland, Malta, Slovakia, and Germany. Cluster 2: Developing insurance markets with low-to-middle IP level and increasing IP pattern until 2010 and varying (decreasing) thereafter. This cluster of countries consists of Cyprus*, Turkey, Greece, and Luxemburg. Cluster 3: Developed insurance markets with middle-to-high IP levels and increasing IP patterns in the whole period. This cluster unites Finland*, Italy, Spain, Denmark, and the Netherlands. Cluster 4: Developing insurance markets with low-to-middle IP levels and increasing IP pattern until 2009 and decreasing thereafter. The within-cluster countries are Croatia*, Slovenia, Iceland, the Czech Republic, Sweden, and Romania. Cluster 5: Developing insurance markets with low-to-middle IP levels and almost quadratic IP

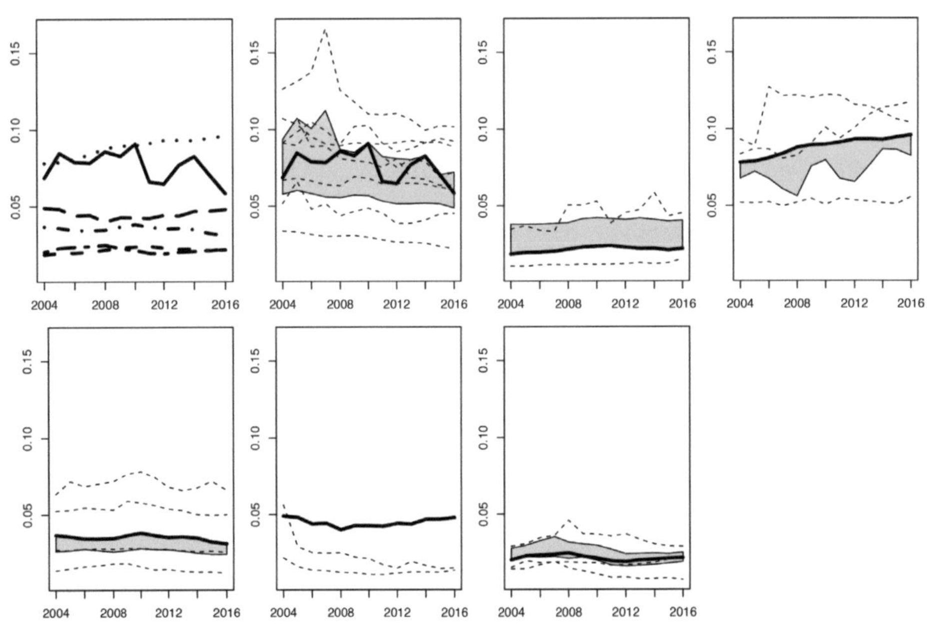

Figure 4.
Clustering results of the IP curves: Magnitude plot.

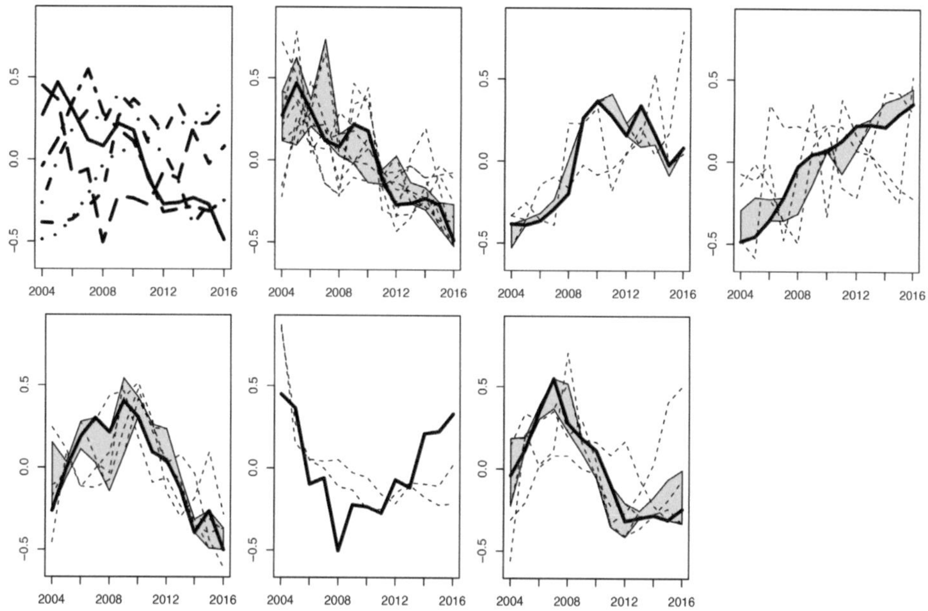

Figure 5.
Clustering results of the IP curves: Shape plot.

pattern and vertex point in 2008. In this cluster, we see countries such as Russia*, Ukraine, and Norway. Cluster 6: Least Developed insurance markets with low IP level and increasing IP pattern followed by a decreasing one initiated in 2007, right on the start of the financial crisis. Members of this cluster are countries such as Lithuania*, Bulgaria, Hungary, Estonia, Serbia, and Poland. The * symbol denotes the medoid IP curve produced by the clustering for each cluster.

The results bring to surface first the difficulty of the European insurance industry to converge and to exhibit homogeneity among national insurance markets during the whole period. A fact that otherwise could lead to the building of single European insurance industry. Second and final, the differential behavior of European insurance markets under different phases of the macroeconomic environment. For instance, Least Developed non-EU insurance markets faced shrink challenges, especially during and after the financial and sovereign debt crises period. The same challenge with a time lag of approximately two years was obvious for some Developing insurance markets. Russia and Ukraine had their insurance markets running in parallel and separated from the other two Developing insurance markets to follow their own smile-shaped development pattern. A slight improvement in insurance activity was also observed for the remaining Developing insurance markets that lasted almost until the end of the sovereign debt crisis in 2011. However, this improvement was offset by their unstable development pattern thereafter. Over the past years, the overall development of Developed insurance markets has decreased, due to a contraction in life insurance business. However, few of them managed to succeed in an increasing pattern with varying IP rate changes over the years.

5. Clustering of population growth data

Over the last century, the world has seen rapid population growth. Particularly, the global population more than quadrupled. The magnitude of the population rate of change from one year to another is found by the fold change ratio (FCR). Fold change is calculated simply as the ratio of the year-end over the year-start population of a certain country. We refer to the evolution of FCR over the course of time as the population growth rate (PGF) curve. In this example, our objective is to find clusters of world countries in which their PGF curves share similar magnitude and shape properties. We use the output of the FC method based on Area measure for clustering world countries. This output will also give a hint towards the distribution of the world population and provide the trends or the dynamics that are defining our world, such that policymakers can set sustainable development goals for our societies.

Thus, we consider the world population data (United Nations 2016), which was analyzed by [21]. This dataset includes estimates of the total population (both sexes) in 233 countries, areas, or regions in July 1950–2015. Motivated by these estimates and the arguments needed for the execution of the FC method, we follow three steps. In the first step, we perform the preprocessing of the dataset by selecting those countries with populations of more than one million in July 1950. In total, 134 countries are included in our analysis. For each of these countries, we collect 65 data points that correspond to the FCR of each year interval and propose connecting them to make the PGF curve. In the second step, we derive the shape information from the L2 normalization of the shifted PGF curves towards their center. This particular step is the one that provides the set of PGF pattern (PGFP) curves. In the last step, we specify the input argument for the number of clusters which is required by FC method to start. The optimal number of clusters was arrived at by calculation of the median value of all methods presented in NBclust library of the R software. Based on the result of this calculation, the chosen number of clusters was three.

Figure 6 satisfies the marginal plot style approach followed in our case studies by presenting the output of the FC method in a two-panel display. The first panel is dedicated to magnitude clustering (it helps discern broad trends in PGF curves), and the second to the shape clustering (it helps identify patterns of pace for population rate of change). The first plot of each panel is the plot of the median curves of the clusters. Remark that the fraction of combinations of countries satisfying the triangular inequality with Area measure was 1 with respect to all combinations.

Next, we present both the derived clusters and their characterization, which is based on the United Nations (UN) geographical region and classification of economies. For instance, we see that the population growth rates in Cluster 1 appear to follow an increasing trend or at least maintain a certain degree of stability because of a natural increase and migration. Most countries in this cluster have a developing economy and are mainly located in Sub-Saharan Africa. However, three European countries (Ireland, Norway and Spain) with developed economies are also members of this cluster of countries. In contrast, the other two characteristic population growth trends that are present in both Clusters 2 and 3 paint a picture of a stagnating or shrinking population in the future, the only difference being that the population in Cluster 3 has a faster speed of shrinkage than in Cluster 2. The most populated cluster (that is Cluster 2 with 64 curves) is mostly associated with another set of developing economies (such as those of Brazil, China and Singapore) located, this time, in Latin America and the Caribbean along with East Asia and Pacific. Additionally, the only developed economy that appears to reside in this cluster is that of the United States, while few economies in transition that belong to the Commonwealth of Independent States (such as those of Azerbaijan, Kazakhstan) make their presence visible for a first time.

Finally, Cluster 3 has united mostly the developed economies of Europe and East Asia and Pacific along with the economies in transition of South-Eastern Europe (Albania, Serbia and North Macedonia). Moreover, the population of few

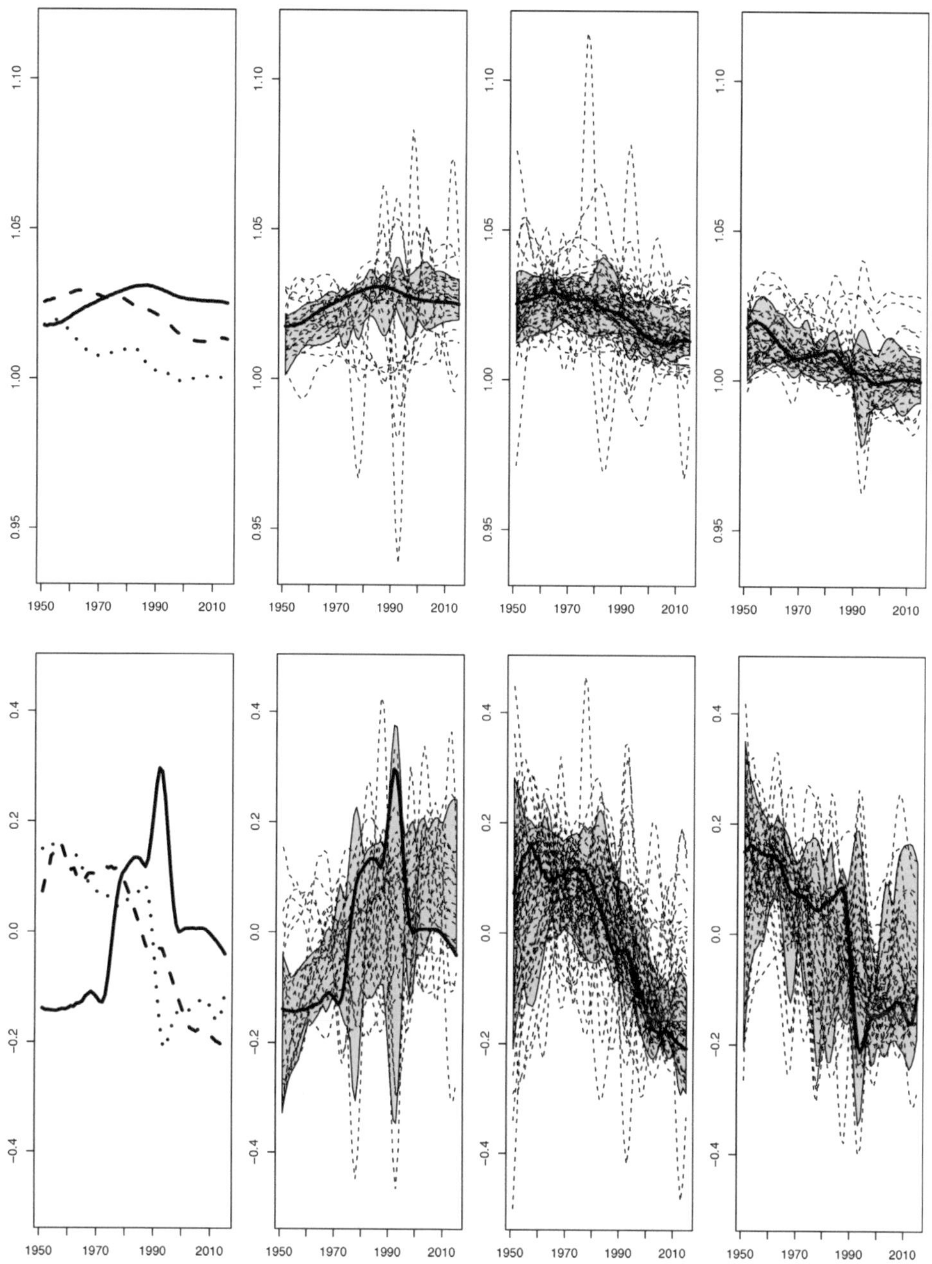

Figure 6.
Clustering results for the population curves. Top panel: Magnitude plot; bottom panel: Shape plot.

developing economies that are located, for example, at Cuba, Jamaica, Puerto Rico, Ghana and Mozambique, have distinguished themselves from the vast majority of developing economies in Cluster 1 or Cluster 2 by following the population behavior of developed economies.

In conclusion, developing economies and economies in transition are split between two clusters, while the majority of the developed economies belong to one cluster. Based on the characterization of these clusters, it is understood that countries with developing economies experience population growth (or at least population stability). However, the more the economy of a country is developed, the more its population growth change decreases. This decrease, in certain cases, might have even a severe negative effect on a country's future projected population size. Whereas, in some other cases, the effect of this decrease is smoother without forcing the population size to reach record lows.

6. Multivariate clustering of insurance penetration with ratio of life and total insurance

The insurance industry generates a large volume of multivariate functional data from the simultaneously obtained measurements on variables related to life, non-life, and total insurance activities. In our case of interest, two main country-specific variables that include data on premiums are available. The first is the total IP (TIP) that represents the development of total activities. While the second is the R ratio of life IP to TIP that represents the development of the share of life premiums in total premiums.

Since the insurance industry of a country can be represented by the bivariate variables of TIP and R, it is important to take into account the dependence between them. We compute a variable that describes this dependence through the covariance function:

$$\mathrm{Cov}(t) = sign((IP(t) - m_1(t))(R(t) - m_2(t)))\sqrt{|(IP(t) - m_1(t))(R(t) - m_2(t))|},$$

where $m_1(t)$ is the mean IP over all countries and $m_2(t)$ is the mean R over all countries and represents the development of the link between total and life share dynamics.

There is no doubt that the development of total activities is different from that of life share. Nevertheless, it may be assumed that a common development coordinates these differential developments of different insurance variables. Then, it is of great interest to identify groups of insurance markets with similar joint development patterns. With this consideration in mind, we aim to discover whether the European insurance market is homogeneous when national insurance developments are jointly differential by developing their total activities and their life share.

We obtained again insurance data from Swiss Re (2016) database and for the same European (EU and non-EU) countries as in univariate case. In particular, we employ a dataset of our main variables for 34 European countries sampled at annual frequency between 2004 and 2016. That is to say that the data for each variable can be viewed as curves. Yet, except for the curves related to TIP and R variables, we also included the computed curves for the Cov variable in our dataset and ended up with a set of three-dimensional vectors of curves.

Viewing the curves for each variable as a set of curves, a three-component list of curve sets is constructed to serve as an input for the FC method. This time, the optimal number of clusters is three and consistent with the median value of all methods presented in the NBclust library of the R software. Our proposed method concentrates on visualizing, in the marginal plot style approach, clusters of multivariate insurance functional data with regard to their magnitudes and covariance function (**Figures 7–9**).

The clustering results are summarized as follows:

Cluster 1: Countries of high TIP and high R with no correlation whatsoever between the two variables throughout the study period.

Cluster 2: Countries of high TIP and high R with a positive correlation between the two variables throughout the study period.

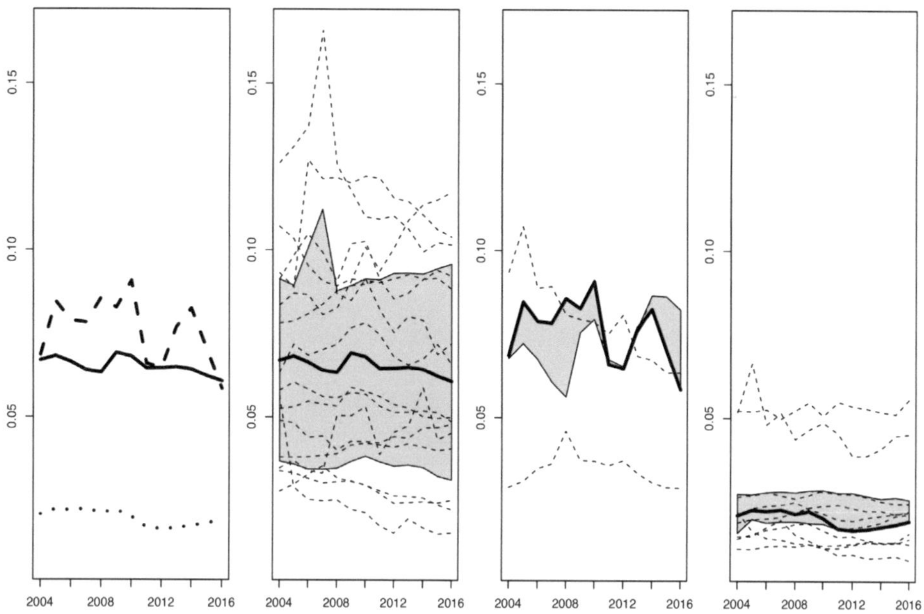

Figure 7.
Clustering results for bivariate curves of TIP and R: TIP plots.

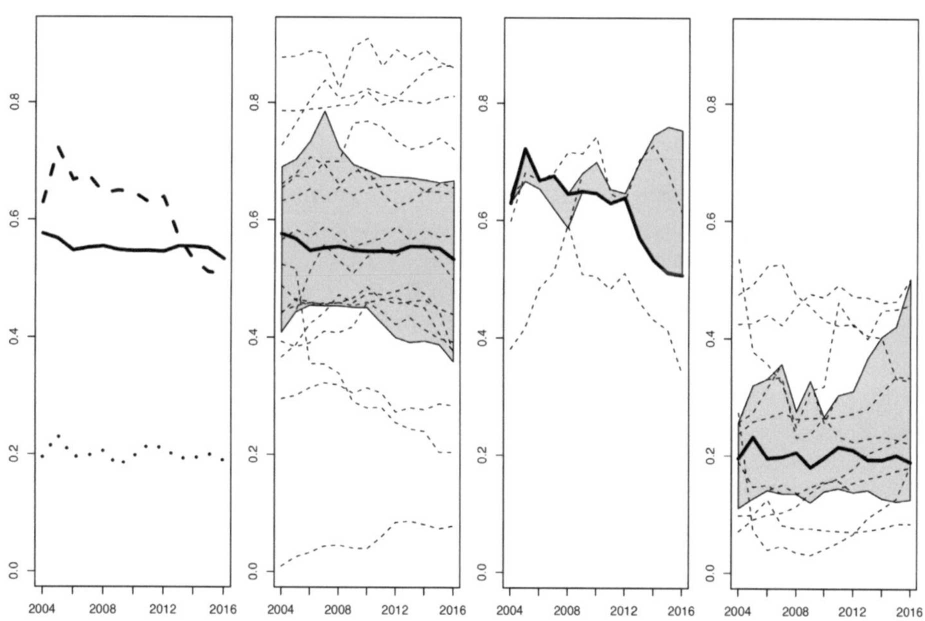

Figure 8.
Clustering results for bivariate curves of TIP and R: R plots.

Cluster 3: Countries of low TIP and low R with no correlation whatsoever between the two variables throughout the study period.

Additionally, the FC method suggests that the total and life share developments in Cluster 1 and Cluster 3 have independent paths since curves for Cov variable almost coincide with the x-axis of **Figure 9**. Simultaneously, it succeeded not to clustered them together due to different magnitude levels. On the contrary, in

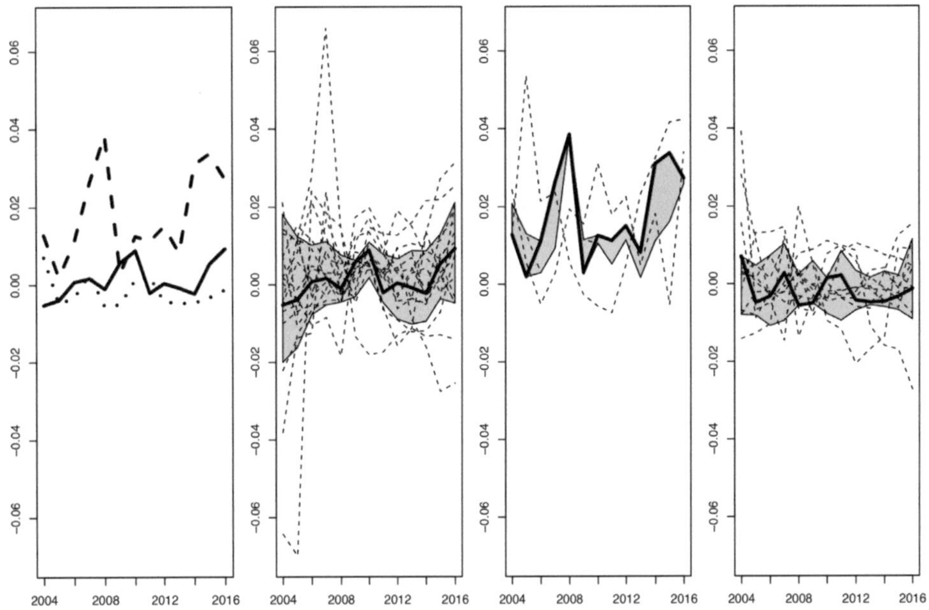

Figure 9.
Clustering results for bivariate curves of TIP and R: Covariance plots.

Cluster 2, the curves for Cov variable are positioned above x-axis, which means that total and life share are dependent functions (positively correlated) over the years.

The functional cluster analysis revealed some differences in the dynamics of insurance markets in Europe. The clustering results clearly reject the hypothesis on the homogeneity of the European insurance market. Europe continues in a two-speed insurance market, with countries with high development and independent paths of total and life insurance business, and others with low. For both speed markets, detecting an increasing pattern in total insurance business does not guarantee that the life premiums will also follow at the same time the same pattern. Any similarity in their patterns could be explained by socio, economic, demographic, or other factors and not by the total business pattern itself. However, there is another high-speed market where the increase of total insurance business in the economy is an additional factor that accelerates the development of life business share.

7. Conclusions

In this study, we introduce a new class of functional cluster analysis methods based on functional orderings. We intended to work with methods that allow intrinsic graphical interpretation to obtain a natural interpretation of clusters via their central regions. Therefore, we propose the use of a studentized measure that forms a metric on the set of functional differences. Also, We suggest the use of the area measure, which orders the functions according to the area of the most extreme continuous rank and considers the entire distribution of the functions. This measure does not form a metric on any set of functions, but the simulation study results and the real data study suggest that it is a metric on any real data set of functions. The check for the satisfaction of the triangular inequality is provided for the given set of functions.

This study's primary aim is to introduce methods that combine the various functional information sources equally. It is possible to study clustering while showing equal concern for both magnitude and shape, as shown in the first and second data examples. In other words, it is possible to study the clustering of multivariate functions when the marginal functions are taken equally. It is also possible to add to the study term, which summarizes the covariance between the marginals of the multivariate function, as shown in the third data example.

The simulation study suggests that the proposed method is robust and more powerful than studied alternatives that give equal treatment to various sources. The studied alternatives are the K-means method, with pre-standardization of every coordinate by its mean and variance, and the model-based method, which assumes a normal distribution of data and considers marginals means, variances, and the covariance function.

Our proposed methods consider the covariance structure of the functional data via the ordering of the entire functional differences. Our proposed methods are also nonparametric and, as such, have no model requirement. Our simulation study also showed that our proposed methods are quite robust to heavy-tailed functions, which can be considered as a type of functional cluster outlier. The data studies show that our methods can cluster the functions with respect to magnitude and shape and that it provides a sensible graphical interpretation of the resulting clusters. The third example shows that the clusters can be also constructed with respect to the covariance of the marginals in the multivariate function. This study does not examine methods to choose the number of clusters in an optimal manner, and this problem is left to the user's choice or further development.

Acknowledgements

Wenlin Dai has been financially supported by National Natural Science Foundation of China (Project No. 11901573). T. Mrkvička has been financially supported by the Grant Agency of Czech Republic (Project No. 19-04412S).

Author details

Wenlin Dai[1], Stavros Athanasiadis[2] and Tomáš Mrkvička[2]*

1 Center for Applied Statistics, Institute of Statistics and Big Data, Renmin University of China, Beijing, China

2 University of South Bohemia, České Budějovice, Czech Republic

*Address all correspondence to: mrkvicka.toma@gmail.com

References

[1] Ieva, F., A. M. Paganoni, D. Pigoli, and V. Vitelli (2013). Multivariate functional clustering for the morphological analysis of electrocardiograph curves. Journal of the Royal Statistical Society: Series C 62 (3), 401–418.

[2] Kazor, K. and A. S. Hering (2015). Assessing the performance of model-based clustering methods in multivariate time series with application to identifying regional wind regimes. Journal of Agricultural, Biological, and Environmental Statistics 20(2), 192–217.

[3] Tupper, L. L., D. S. Matteson, C. L. Anderson, and L. Zephyr (2018). Band depth clustering for nonstationary time series and wind speed behavior. Technometrics 60(2), 245–254.

[4] Ferraty, F. and P. Vieu (2006). Nonparametric Functional Data Analysis: Theory and Practice (1 ed.). Springer Series in Statistics. Springer.

[5] Chiou, J.-M. and P.-L. Li (2007). Functional clustering and identifying substructures of longitudinal data. Journal of the Royal Statistical Society: Series B 69(4), 679–699.

[6] Jacques, J. and C. Preda (2014). Model-based clustering for multivariate functional data. Computational Statistics and Data Analysis 71, 92–106.

[7] Zeng, P., J. Q. Shi, and W.-S. Kim (2019). Simultaneous registration and clustering for multidimensional functional data. Journal of Computational and Graphical Statistics 28(4), 943–953.

[8] López-Pintado, S. and J. Romo (2009). On the concept of depth for functional data. Journal of the American Statistical Association 104(486), 718–734.

[9] López-Pintado, S., Y. Sun, J. K. Lin, and M. G. Genton (2014). Simplicial band depth for multivariate functional data. Advances in Data Analysis and Classification 8(3), 321–338.

[10] Myllymäki, M. and T. Mrkvička (2020). Get: Global envelopes in r. arXiv preprint arXiv:1911.06583.

[11] Sguera, C., P. Galeano, and R. Lillo (2014). Spatial depth-based classification for functional data. TEST 23(4), 725–750.

[12] de Micheaux, P. L., P. Mozharovskyi, and M. Vimond (2020). Depth for curve data and applications. Journal of the American Statistical Association 0(0), 1–17.

[13] Dai, W., T. Mrkvička, Y. Sun, and M. G. Genton (2020). Functional outlier detection and taxonomy by sequential transformations. Computational Statistics and Data Analysis 149, 106960.

[14] Myllymäki, M., P. Grabarnik, H. Seijo, and D. Stoyan (2015). Deviation test construction and power comparison for marked spatial point patterns. Spatial Statistics 11, 19–34.

[15] Myllymäki, M., T. Mrkvička, P. Grabarnik, H. Seijo, and U. Hahn (2017). Global envelope tests for spatial processes. J. R. Statist. Soc. B 79(2), 381–404.

[16] Narisetty, N. N. and V. N. Nair (2016). Extremal depth for functional data and applications. Journal of American Statistical Association 111 (516), 1705–1714.

[17] Febrero-Bande, M. and M. Oviedo de la Fuente (2012). Statistical computing in functional data analysis: The R package fda.usc. Journal of Statistical Software 51(4), 1–28.

[18] Fraley, C. and A. E. Raftery (2002). Model-based clustering, discriminant

analysis, and density estimation. Journal of the American Statistical Association 97(458), 611–631.

[19] Carroll, C., A. Gajardo, Y. Chen, X. Dai, J. Fan, P. Z. Hadjipantelis, K. Han, H. Ji, H.-G. Mueller, and J.-L. Wang (2021). fdapace: Functional Data Analysis and Empirical Dynamics. R package version 0.5.6.

[20] Hubert, L. and P. Arabie (1985). Comparing partitions. Journal of classification 2(1), 193–218.

[21] Nagy, S., I. Gijbels, and D. Hlubinka (2017). Depth-based recognition of shape outlying functions. Journal of Computational and Graphical Statistics 26(4), 883–893.

Chapter 2

Computational Statistics with Dummy Variables

Adji Achmad Rinaldo Fernandes, Solimun and Nurjannah

Abstract

Cluster analysis is a technique commonly used to group objects and then further analysis is carried out to obtain a model, named cluster integration. This process can be continued with various analyzes, including path analyzes, discriminant analyzes, logistics, etc. In this chapter, the author discusses the reason to use dummy variables in this type of cluster analysis. Dummy variables are the main way that categorical variables are included as predictors in modeling. With statistical models such as linear regression, one of the dummy variables needs to be excluded, otherwise the predictor variables are perfectly correlated. Thus, usually if a categorical variable can take *k* values, we only need *k-1* dummy variables, the *k-th* variable being redundant, it does not bring any new information. When more dummy variables than needed are used this is known as dummy variable trapping. The advantage to use dummy variables is that they are simple to use and the decision making process is easier to manage. The novelty in this chapter is the perspective of the dummy variable technique using cluster analysis in statistical modeling. The data used in this study is an assessment of the provision of credit risk at a bank in Indonesia. All analyzes were carried out using software R.

Keywords: dummy, cluster, integrated cluster with logistic regression, integrated cluster with discriminant analysis, integrated cluster with path analysis

1. Introduction

The application of cluster analysis is commonly used to group objects. Cluster analysis can be used to group objects and then further analysis is carried out to obtain a model, namely cluster integration. Cluster integration can be continued with various analyzes, including path analysis, discriminant analysis, logistics, etc. In cluster integration with path analysis, it aims to group homogeneous objects into one group, the goal is that the resulting residual variance is homogeneous in addition to maximizing the adjusted R^2 value. In cluster integration with discriminant analysis, the benefits of cluster analysis generated can maximize the accuracy, sensitivity, and specificity of the model. In this chapter, we will explain the technical perspective of dummy variables using cluster analysis in statistical modeling, such as regression analysis, path analysis, and discriminant analysis.

2. Why use dummy variables

Dummy variables are numerical variables that represent categorical data, such as gender, race, political affiliation, etc. Technically, the dummy variable is dichotomous, a quantitative variable. Their value range is small; they can only take two quantitative values. As a practical matter, regression results are easiest to interpret when the dummy variable is constrained to two specific values, 1 or 0. Typically, 1 represents the presence of a qualitative attribute, and 0 represents its absence. Categorical variables have more than two categories, can be represented by a set of dummy variables, with one variable for each category. Numerical variables can also be coded to explore nonlinear effects. Dummy variables are also known as indicator variables, design variables, contrasts, one-hot coding, and binary basis variables [1].

Dummy variables are the main way that categorical variables are included as predictors in modeling. For example, in linear regression analysis, the response variable is profit, and the predictor variable is employee group. With statistical models such as linear regression, one of the dummy variables needs to be excluded (by convention, the former, or the latter), otherwise, the predictor variables are perfectly correlated [2].

When defining dummy variables, a common mistake is to define too many variables. If a categorical variable can take on *k* values, then you tend to define *k* dummy variables. You only need *k-1* dummy variable.

The *k-th* dummy variable is redundant; it does not bring any new information. And that creates a severe multicollinearity problem for analysis. Using *k* dummy variables when only *k-1* dummy variables are needed is known as dummy variable trapping.

Regression analysis treats all independent variables (X) in the analysis as numerical. A numeric variable is an interval or ratio scale variable whose values can be directly compared, e.g. "10 is double 5," or "3 minus 1 equals 2." However, you may want to include a nominal scale attribute or variable such as: "Product Brand" or "Defect Type" in your study. Say you have three types of defects, numbered "1," "2" and "3." In this case, "3 minus 1" means nothing. You cannot subtract handicap 1 from handicap 3. The numbers here are used to indicate or identify the degree of "Type of Disability" and have no intrinsic meaning of their own. A dummy variable is created in this situation to "trick" the regression algorithm into the correct attribution of the analysis variable [3].

The main benefit of dummy variables is that they are simple. Often there are better alternative basis functions, such as orthogonal polynomials, effect coding, and splines. If dummy variables are used in linear regression analysis, then there are several advantages [4], including:

a. The dependent variable prediction process becomes more focused and accurate, different from ordinary multiple regression

b. Because the data is not qualitative, the prediction results are easy to interpret

c. The decision-making process tends to be easy

3. Hierarchical cluster

Cluster analysis (group analysis) is an analytical method that aims to group objects into several groups, objects in groups are homogeneous (same) while other group members are heterogeneous (different) [5]. The procedure for group formation in Cluster analysis is divided into two, namely hierarchical and non-hierarchical methods. Grouping with the hierarchical method is used when there is no information about the number of clusters. The main principle of the hierarchical method is to group objects that have something in common with one group. While the non-hierarchical method is used when information about the number of clusters is known or has been determined [6].

This method starts grouping with two or more objects that have the closest object. Then the process is continued by passing to another object that has second proximity. And so on to form a tree in which there is a hierarchy or level from the most similar to the different. The tree formed by this cluster is also called a dendrogram. This tree is useful for providing deeper clarity on the clustering process.

The stages of grouping data using the hierarchical method are [7]:

1. Determine k as the number of clusters to be formed.

2. Each object data is considered as a cluster so that n = N.

3. Calculate the distance between clusters.

4. Find two clusters that have the least distance between clusters and combine them (meaning N = n-1).

5. If n > k, then go back to step 3.

According to [6] in the method of forming groups in the hierarchical method, there are two approaches, namely agglomerative hierarchical methods (Agglomerative Hierarchical Methods) and divisive hierarchical methods (Device Hierarchical Methods). The agglomerative method starts by assuming that each object is a cluster. Then the two objects that have the closest distance are combined into one cluster. The process continues so that in the end it will form a cluster consisting of all objects.

4. Integrated cluster with logistic regression

4.1 Integrated cluster equation model with logistic regression analysis

The model of the integration of cluster analysis with logistic analysis of the dummy variable approach is the same as the general model of multiple linear regression analysis with dummy variables.

The general model of the integrated cluster with logistic analysis can be written in the following Eq. (1).

$$y_i = \frac{\exp\begin{pmatrix} \beta_0 + \beta_1 x_{1i} + \cdots + \beta_p x_{pi} + \\ D_1\beta_{p+1} + D_1\beta_{p+1}x_{1i} + \cdots + D_1\beta_{2p+1}x_{pi} + \\ D_2\beta_{p+2} + D_2\beta_{p+3}x_{1i} + \cdots + D_2\beta_{3p+2}x_{pi} + \\ \cdots + D_q\beta_{pq+1} + D_q\beta_{pq+2}x_{1i} + \cdots + D_q\beta_{p(q+1)}x_{pi} \end{pmatrix}}{1 + \exp\begin{pmatrix} \beta_0 + \beta_1 x_{1i} + \cdots + \beta_p x_{pi} + \\ D_1\beta_{p+1} + D_1\beta_{p+1}x_{1i} + \cdots + D_1\beta_{2p+1}x_{pi} + \\ D_2\beta_{p+2} + D_2\beta_{p+3}x_{1i} + \cdots + D_2\beta_{3p+2}x_{pi} + \\ \cdots + D_q\beta_{pq+1} + D_q\beta_{pq+2}x_{1i} + \cdots + D_q\beta_{p(q+1)}x_{pi} \end{pmatrix}} \tag{1}$$

where,
y_i: response variable at the *i-th* observation unit
x_{ki}: the *k-th* predictor variable on the *i-th* observation unit
β_p: coefficient of the *p-th* logistic function
D_q: *q-th* dummy variable
p: number of predictor variables
q: the number of clusters formed is reduced by 1
i: 1, 2, 3, … , n

4.2 Logistics regression analysis assumptions

Before conducting the analysis, several basic principles or assumptions underlie regression analysis, several assumptions that underlie logistic regression analysis, namely [8].

1. Does not assume a linear relationship between the response variables and the predictor variables.

2. Predictor variables do not have to be normally distributed.

3. The response variable does not require the assumption of homogeneity for each level of the predictor variable or the variance does not have to be the same in each category.

4. The measurement scale on the response variable is discrete or binary (success/failure) and the predictor variable does not require an interval measurement scale.

5. Using probability sampling, which is a sampling technique to provide equal opportunities for each member of the population to be selected as a member of the sample.

6. Observation variables are measured without errors (valid and reliable measurement instruments) meaning that the variables studied can be observed directly.

4.3 Integrated cluster analysis method with logistic regression analysis

The linkage used in this study is the Average Linkage and the measurement of the distance between clusters using the Euclidean distance. Determination of the number has been determined in advance, namely as many as 2 and 3 groups. The Average Linkage method is based on the average distance. The table of the number of members in each Cluster in the Integrated Cluster Analysis method with regression analysis is presented in **Table 1**.

From **Table 1** it can be seen that there are 71 customers in Cluster 1 with 3 groups, 15 customers in Cluster 2, and 14 customers in Cluster 3. While many members with 2 groups in Cluster 1 as many as 93 customers and in Cluster 2 as many as 7 customers. The selection of the best linkage and model validity is by choosing the model that has the largest total R^2, as shown in the equation, which can be briefly seen in **Table 2** as follows.

Based on **Table 2** the logistic regression analysis model with cluster integration with 3 groups has the greatest total determination value so that logistic regression

Cluster	Number of cluster members	
	3 Groups	2 Groups
1	71	93
2	15	7
3	14	—

Table 1.
Number of members of each cluster average linkage method on integrated cluster analysis method with logistics regression analysis.

	R^2 adjusted of Y_1	R^2 adjusted of Y_2	Total R^2 adjusted
3 Groups	0.4258	0.8492	0.8923
2 Groups	0.3852	0.8129	0.8667

Table 2.
Adjusted values R^2 for each integrated cluster analysis model with logistics regression analysis.

analysis with cluster integration with 3 groups is the best model compared to 2 groups. The total determination value of 89.23% is considered very good to describe the model.

Based on **Table 2** the adjusted R^2 value of the Cluster integration regression analysis with 3 groups resulted in an adjusted R^2 value of 0.4258 meaning that the variables of age, work experience, and loan to value were able to explain the diversity of credit collectibility variables of 42.58%, while 57 The other, 41% is influenced by variables outside the model. The value of R^2 adjusted Cluster integration logistic regression analysis with 3 groups resulted in an R^2 adjusted value of 0.8492, meaning that the variables of age, work experience, and loan to value were able to explain the diversity of credit collectibility variables of 84.92%, while 13.08. The other percentage is influenced by variables outside the model. The coefficient of total determination of the Cluster integration logistic regression analysis model with 3 groups is 0.8923, so it can be concluded that the diversity of data that can be explained by the model is 89.23% while the remaining 10.17% is explained by variables outside the model.

The results of R^2 the adjusted integrated cluster in logistic regression analysis with 3 groups having the highest adjusted R^2 value. If the average variables of each Cluster are compared, it is found that most of Cluster 2 has the highest average value compared to other Clusters, so Cluster 2 is high. While Cluster 1 has the lowest average value compared to other Clusters, so Cluster 1 is low. The average value for each cluster is presented in **Table 3**.

Based on **Table 3**, it can be seen that most of the customers are 39 years old in the low cluster, 37 years old in the high cluster, and 38 years old in the medium

Variable	Average		
	Cluster 1: low cluster	Cluster 2: high cluster	Cluster 3: medium cluster
Age (X_1)	39.507	37.333	38.571
Work experience (X_2)	39.930	193.867	107.571

Table 3.
Average value and each cluster in integrated cluster analysis model with logistic regression analysis.

cluster. The work experience of customers in the low cluster is mostly for 40 months, the high cluster is mostly for 194 months, while in the medium cluster mostly for 108 months.

Integrated Cluster Analysis method with Logistic Regression Analysis with 3 groups that separate each data set optimally. Then the model formed is like Eq. (2) as follows.

$$\pi(x) = \frac{\exp\left(-0,027x_1 - 0,041x_2 + 0,850y_1 - 0,374D_1x_1 - 0,006D_1x_2 + 9,971D_1y_1 + 0,090D_2x_1 + 0,026D_2x_2 - 1,559D_2y_1\right)}{1 + \exp\left(-0,027x_1 - 0,041x_2 + 0,850y_1 - 0,374D_1x_1 - 0,006D_1x_2 + 9,971D_1y_1 + 0,090D_2x_1 + 0,026D_2x_2 - 1,559D_2y_1\right)} \tag{2}$$

Low cluster ($D_1 = 0$ and $D_2 = 0$) can be seen in Eq. (3).

$$\pi(x) = \frac{\exp\left(-0,027x_1 - 0,041x_2 + 0,850y_1\right)}{1 + \exp\left(-0,027x_1 - 0,041x_2 + 0,850y_1\right)} \tag{3}$$

High cluster ($D_1 = 1$ and $D_2 = 0$) can be seen in Eq. (4).

$$\pi(x) = \frac{\exp\left(-0,401x_1 - 0,047x_2 + 10,821y_1\right)}{1 + \exp\left(-0,401x_1 - 0,047x_2 + 10,821y_1\right)} \tag{4}$$

Medium cluster ($D_1 = 0$ and $D_2 = 1$) can be seen in Eq. (5).

$$\pi(x) = \frac{\exp\left(0,063x_1 - 0,015x_2 - 0,709y_1\right)}{1 + \exp\left(0,063x_1 - 0,015x_2 - 0,709y_1\right)} \tag{5}$$

5. Integrated cluster with discriminant analysis

5.1 Discriminant analysis

Discriminant analysis is a multivariate analysis that functions to model the relationship between a categorical response variable and one or more quantitative predictor variables [9]. Discriminant analysis can be used as a grouping method because it produces a function that cancan distinguishes between groups. The function is formed by maximizing the distance between groups. If the response variable or categorical data consists of only two groups, it is called a Two-Group Discriminant Analysis model, whereas if the group consists of more than two categories it is called Multiple Discriminant Analysis. Discriminant analysis has two assumptions that must be met, namely the assumption of multivariate normality, and the assumption of homogeneity of the variance matrix.

According to [6], discriminant analysis is included in the multivariate dependence method. The model can be written as in Eq. (6).

$$y_i = \beta_1 X_{1i} + \beta_2 X_{2i} + \ldots + \beta_p X_{pi} \tag{6}$$

where,

y_i: the response variable is categorical or nominal data on the *i-th* observation unit

X_{pi}: the p-explanatory variable on the *i-th* observation unit

β_p: the coefficient of the *p-th* discriminant function

i: 1, 2, 3, …, n

5.2 Integration of cluster analysis with discriminant analysis of dummy variable approach

Integration of Cluster Analysis with Discriminant Analysis The Dummy Variable Approach in this study combines cluster analysis with discriminant analysis. Integrating cluster analysis with discriminant analysis can be done by using dummy variables obtained from cluster results. Many clusters formed are used as categories, then used as dummy variables.

An integrated cluster model with discriminant analysis can be written in Eq. (7).

$$\begin{aligned} y_i = &\beta_1 x_{1i} + \beta_2 x_{2i} + \dots + \beta_p x_{pi} \\ &+ D_1\beta_{p+1}x_{1i} + D_1\beta_{p+2}x_{2i} + \dots + D_1\beta_{p+q}x_{pi} \\ &+ D_2\beta_{p+q+1}x_{1i} + D_2\beta_{p+q+2}x_{2i} + \dots + D_2\beta_{p+2q}x_{pi} \\ &+ \dots + D_q\beta_{p+qq+1}x_{1i} + D_q\beta_{p+11+2}x_{2i} + \dots + D_q\beta_{p+qq}x_{pi} \end{aligned} \tag{7}$$

where,
y_i: response variable at the *i-th* observation unit
x_{pi}: the p-explanatory variable on the *i-th* observation unit
β_p: the coefficient of the *p-th* discriminant function
D_q: *q-th* dummy variable
p: number of explanatory variables
q: the number of clusters formed is reduced by 1
i: 1, 2, 3, …, n

If the research variables used are 3 and the number of clusters is 2, then the integrated cluster model with multiple discriminant analysis can be written as in Eq. (8).

Common models:

$$y_i = \beta_1 x_{1i} + \beta_2 x_{2i} + \beta_3 x_{3i} + D_1\beta_4 x_{1i} + D_1\beta_5 x_{2i} + D_1\beta_6 x_{3i} \tag{8}$$

Cluster 1 (D_1=0)

$$y_i = \beta_1 x_{1i} + \beta_2 x_{2i} + \beta_3 x_{3i} \tag{9}$$

Cluster 2 (D_1=1)

$$y_i = (\beta_1 + \beta_4)x_{1i} + (\beta_2 + \beta_5)x_{2i} + (\beta_3 + \beta_6)x_{3i} \tag{10}$$

5.3 Model efficiency

Efficiency can be seen based on three criteria, namely model accuracy, sensitivity, and specificity. Accuracy measures how correctly a diagnostic test identifies and excludes a certain condition, in other words, accuracy is used to measure the goodness of the model. In diagnostic tests, the terms sensitivity and specificity are also known. Sensitivity and specificity in diagnostic tests is a measure of the ability to correctly identify objects under reality [10]. The difference is that sensitivity measures the positive group while specificity measures the negative group. To get the value of accuracy, sensitivity, and specificity can use the Confusion Matrix as follows (**Table 4**).

$$\begin{aligned} \text{Accuracy} &= \frac{a+d}{a+b+c+d} \\ \text{Sensitivity} &= \frac{a}{a+c} \\ \text{Specificity} &= \frac{d}{b+d} \end{aligned} \tag{11}$$

Actual	Prediction	
	Z_1	Z_0
Z_1	a	b
Z_0	c	d

Table 4.
Confusion matrix.

5.4 Implementation of integrated cluster with discriminant analysis

For example, there are secondary data regarding homeownership loans obtained from Bank X in Indonesia, where the variables studied are age, credit period, loan to value, and credit collectibility status. The collectibility status of the credit used consists of two categories, namely the collectibility of current and non-current loans. The variables of age and credit period are in hours, while the loan to value is in proportion units. Therefore, it is necessary to standardize before conducting data analysis.

When using an integrated cluster with discriminant analysis, the first thing we have to do is perform a cluster analysis to get a dummy variable. In cluster analysis, it is not necessary to test assumptions because cluster analysis is included in exploratory analysis. If the results of the cluster analysis are *n* clusters, then the dummy variables formed are *n-1* variables. The analysis used is hierarchical cluster analysis with the average linkage method with Euclidean distance. The determination of the number of clusters is determined based on the Silhouette value. Silhouette values for each of the many clusters can be seen in **Table 5**.

Based on **Table 5**, the largest Silhouette value is in many clusters 2. So that the optimum number of clusters is 2. The results of cluster analysis are obtained in cluster 1 consisting of 71 customers, and cluster 2 consisting of 29 customers. Thus, the dummy variable formed is 1 dummy variable. If the object (customer) is included in cluster 2, we assume that the object is 1 in the dummy variable. Meanwhile, if the object is included in cluster 1, we assume that the object is 0 in the dummy variable.

After obtaining the dummy variable, the next step is to test the assumptions in discriminant analysis. Testing for multivariate normality using the Shapiro-Wilk test on predictor variables, and testing the homogeneity of the covariance matrix using the Box M test. of 0.9917 (> 0.05). So it can be concluded that the data already meet the assumptions of multivariate normality and homogeneity of the variance matrix.

Next is to analyze the data using an integrated cluster with discriminant analysis. Based on the analysis carried out, an integrated cluster model was obtained with the following discriminant analysis:

Number of clusters	Silhouette value
2	0.4491
3	0.3915
4	0.2912
5	0.2811

Table 5.
Cluster analysis silhouette results.

$$y_i = 0,0838x_{1i} + 0.0606x_{12i} - 0,0241x_{3i} + 0,0569D_1x_{1i} + 0,0358D_1x_{2i} - 0,0752D_1x_{3i} \quad (12)$$

Cluster 1 (D_1 = 0)

$$y_i = 0,0838x_{1i} + 0.0606x_{12i} - 0,0241x_{3i} \quad (13)$$

Cluster 1 (D_1 = 1)

$$y_i = 0,1407x_{1i} + 0.0964x_{12i} - 0,0993x_{3i} \quad (14)$$

Based on the above equations, it can be interpreted that the coefficient of age and credit term is positive, meaning that the higher the age and credit term, the greater the possibility that customers in cluster 1 and cluster 2 have current credit collectability. On the other hand, loan-to-value has a negative coefficient, so if the value increases, it will increase the possibility of customers having non-current credit collectibility. The variable that most influences credit collectibility in cluster 1 and cluster 2 is age which has the largest discriminant coefficient value. The value of classification accuracy, sensitivity, and specificity in the integrated cluster analysis method with discriminant analysis can be seen in **Table 6** below.

Based on **Table 6**, the results of the classification accuracy are 84%, which means that the model correctly classifies as many as 84 customers out of 100 customers. Sensitivity of 84% means that customers belonging to the current category can be classified correctly by the model as many as 60 of 71 customers. The specificity of 16% means that customers belonging to the non-current category can be classified correctly by the model as many as 5 out of 29 customers.

6. Regression analysis with dummy variable

6.1 Regression analysis

The method that describes how big the relationship between variables is a regression analysis method. Regression analysis is divided into two, namely simple regression analysis and multiple regression analysis. Simple regression analysis is an analysis involving one predictor variable and one response variable, while multiple regression analysis is a regression analysis involving several predictor variables and one response variable. The regression analysis has several classical assumptions based on Gauss-Markov theory that must be met, namely the relationship between variables is correct, predictor variables are fixed or non-stochastic, homogeneity of variance, non-autocorrelation, error normality, non-multicollinearity [11].

6.2 Regression analysis with dummy variables

There are many ways to create a regression model with qualitative predictor variables, one of which is to use regression with dummy variables. The dummy variable is a variable used to obtain an estimator in a regression model involving

	Percentage
Classification accuracy	84%
Sensitivity	84%
Specificity	16%

Table 6.
Value of classification accuracy, sensitivity, and specificity.

qualitative predictor variables [12]. There is no difference in the assumptions underlying the regression with or without a dummy variable, this is because the addition of a dummy variable will be the same as the addition of a predictor variable in general.

There are several rules for coding dummy variables, for example by using binary code (0, 1). For example, there is a qualitative predictor variable with two categories (category 1 and category 2), then the qualitative variable can be defined in the dummy variable as shown in the following equation:

$$D = \begin{cases} 1, & \textit{for category } 1 \\ 0, & \textit{for other} \end{cases} \tag{15}$$

The regression model with dummy variables can be expressed in the following equation:

$$Y_i = \beta_0 + \beta_1 D + \beta_2 X_2 + \beta_3 X_3 + \beta_4 X_4 + \varepsilon_i \tag{16}$$

Information:
Y_i: the value of the response variable at the *i-th* observation.
X_i: the value of predictor *i-th* variable.
β: regression model parameter.
ε_i: Random error at *i-th* observation.
i: index for observation $(i = 1, 2, \ldots, n)$.
Dummy variables can be entered into the regression model in three different ways, namely:

1. Dummy variable as intercept component

2. Dummy variable as slope component

3. Dummy variables as components of intercept and slope

6.3 Application of regression analysis with dummy variables

From the available data, namely Y = willingness to pay, X_1 = dummy variable with category 1 being income in one family that is not combined, while category 2 is income in one family combined. X2 is Service Quality, X3 is Environment and X4 is Fairness.

The regression model formed is $Y = b_0 + b_1 D + b_2 X_2 + b_3 X_3 + b_4 X_4$ (**Figure 1**)
Based on the regression analysis performed, the regression model with dummy variables is obtained as follows:

$$Y = 0.54088 + 0.08676\,D + 0.1579X_2 + 0.4309X_3 + 0.2545X_4 \tag{17}$$

In this model, it is possible to know the difference in interest in paying creditors whose income is combined with income that is not combined.

6.4 Assumptions of regression analysis with dummy variable

6.4.1 Non multicollinearity

Multicollinearity is a problem in regression which means that the predictor variables correlate. A good regression model is a data that there is no

```
Coefficients:
            Estimate Std. Error t value Pr(>|t|)
(Intercept)  0.54088    0.23671   2.285 0.024540 *
dat$X11      0.08676    0.08096   1.072 0.286571
dat$X2       0.15794    0.07425   2.127 0.035992 *
dat$X3       0.43093    0.05248   8.211 1.08e-12 ***
dat$X4       0.25451    0.07078   3.596 0.000515 ***
---
Signif. codes:  0 '***' 0.001 '**' 0.01 '*' 0.05 '.' 0.1 ' ' 1

Residual standard error: 0.3269 on 95 degrees of freedom
Multiple R-squared:  0.684,	Adjusted R-squared:  0.6707
F-statistic: 51.42 on 4 and 95 DF,  p-value: < 2.2e-16
```

Figure 1.
Output R.

multicollinearity problem. Multicollinearity checks can use the VIF value, where if the VIF value is <10 then there is no multicollinearity problem. In the data used, the VIF value for all variables is less than 10 so that the data is used to fulfill the assumption of non-multicollinearity.

6.4.2 Normality error

The assumption of normality of error is an assumption that requires that the error must be normally distributed with a mean of 0 and a variance σ^2. Testing for normality of errors can use the Shapiro Wilk test.

H0: normal distribution error

H1: error is not normally distributed

$\alpha = 5\%$

Based on the normality test, a p-value of 0.91 was obtained, which means that the error was normally distributed. So that the assumption of normality error is met.

6.4.3 Non autocorrelation

The non-autocorrelation assumption test aims to find out whether some observations have correlated errors or not. If there is covariance and the correlation between errors is not equal to zero, then this can be said as a violation of assumptions. The non-autocorrelation assumption test method can be done using the Durbin Watson method. Based on the analysis conducted using the Durbin Watson test, a p-value of 0.6132 was obtained, which means that the data met the non-autocorrelation assumption.

6.4.4 Homoscedasticity

The assumption of homoscedasticity invariance indicates that as the average increases, the variance should remain constant, but there is a possibility that an increase in the average value causes the variance value to also increase, so it is necessary to test the assumption of homogeneity of variance. Assumption testing is done so that the estimator results obtained are efficient. Testing the assumption of homoscedasticity can use the Brusch Pagan method.

Based on the analysis conducted using the Brusch Pagan test, a p-value of 0.130 (less than 0.05) was obtained, which means that the data met the assumption of homoscedasticity.

6.5 Parameter significance test

a. Simultaneous test

$H_0: \beta_1 = \beta_2 = \beta_3 = \beta_4 = 0$

$\boldsymbol{H_1}$: there is at least one $\boldsymbol{\beta_i \neq 0}$

$\alpha = 5\%$

Based on the analysis obtained a p-value of 0.000 which means that there is at least one significant regression coefficient.

b. Partial test

$H_0: \beta_i = 0$

$H_1: \beta_i \neq 0$

$\alpha = 5\%$

Based on the analysis, it was found that three regression coefficients have a p-value of less than 0.05. The three regression coefficients are the coefficients of the variables X2 (Quality of Service), X3 (Environment), and X4 (Fairness). This means that Service Quality, Environment, and Fairness have a significant effect on Willingness to Pay.

6.6 Model interpretation

The model obtained and has fulfilled all the assumptions of regression analysis with dummy variables is as follows:

$$y = 0.54088 + 0.08676\, D + 0.1579x_2 + 0.4309x_3 + 0.2545x_4 \tag{18}$$

In this model, it is possible to find out the difference in interest in paying creditors whose income is combined with income that is not combined. Based on the model obtained, the coefficient of the dummy variable is 0.08676, which means that when the incomes of creditors in one family are combined, the willingness to pay will be greater than those of creditors whose income is not combined. The estimated regression coefficient for the variable X2 (Quality of Service) is 0.1579, which means that the better the bank's service quality, the willingness to pay for credit also increases. Then for the estimation of the regression coefficient for the X3 (Environmental) variable, an estimate of 0.4309 is obtained, which means that the better the creditor's environmental conditions, the willingness to pay credit will also increase. The same thing also happened to the variable X4 (Fairness) where the estimated regression coefficient was 0.2545, which means that if the bank institution is fairer, creditors will also be more interested in paying.

7. Conclusion

The use of cluster analysis in statistical modeling will greatly facilitate the capture of the diversity of objects so that objects with the same characteristics can be grouped into the same group. This will be useful in classification methods such as discriminant analysis. Because in one group, objects will be more homogeneous, while between groups has a high diversity. So, the novelty in this chapter is the perspective of the dummy variable technique where the number of categories in the dummy variable is determined by the number of clusters formed from the results of cluster analysis. This will then be continued on statistical modeling which is able to help researchers to divide objects into several groups according to the characteristics of each object by minimizing the diversity within the group.

Conflict of interest

The authors declare no conflict of interest.

Author details

Adji Achmad Rinaldo Fernandes*, Solimun and Nurjannah
Departement of Statistics, University of Brawijaya, Malang City, Indonesia

*Address all correspondence to: fernandes@staff.ub.ac.id

References

[1] Stattrek.com. Dummy Variables in Regression. 2021. Available from: https://stattrek.com/multiple-regression/dummy-variables.aspx

[2] Displayr.com. What are Dummy Variables?. 2019. Available from: https://www.displayr.com/what-are-dummy-variables/

[3] Skrivanek S. The Use of Dummy Variables in Regression Analysis. Powell, OH: More Steam, LLC; 2009

[4] Artaya IP. Analisa Regresi Linier Berganda Metode Dummy Banyak Kriteria. 2019. DOI: 10.13140/RG.2.2.13471.41122

[5] Fernandes AAR. Metode Statistika Multivariat Pemodelan PErsamaan Struktural (SEM) Pendekatan WarPLS. Malang: UB Press; 2017

[6] Johnson RA, Wichern DW. Applied Multivariate. Analysis. Upper Saddle River. NJ: Prentice-Hall; 2007

[7] Gudono. Analisis Data Multivariat Edisi Pertama. Yogyakarta: BPFE; 2011

[8] Tatham RL, Hair JF, Anderson RE, Black WC. Multivariate Data Analysis. New Jersey: Prentice Hall; 1998

[9] Wong HB, Lim GH. Measures of Diagnostic Accuracy: Sensitivity, Specificity, PPV and NPV. Proceedings of Singapore Healthcare. 2011;**20**(4):316-318

[10] Gujarati D. Ekonometri Dasar: Terjemahan Sumarno Zein. Erlangga: Jakarta; 2003

[11] Nawari. Analisis Regresi dengan MS Excel 2007 SPSS 17. Jakarta: PT Elex Media Komputindo; 2010

[12] Le Cessie S, Van Houwelingen JC. Logistic regression for correlated binary data. Journal of the Royal Statistical Society: Series C (Applied Statistics). 1994;**43**(1):95-108

Chapter 3

Sparse Boosting Based Machine Learning Methods for High-Dimensional Data

Mu Yue

Abstract

In high-dimensional data, penalized regression is often used for variable selection and parameter estimation. However, these methods typically require time-consuming cross-validation methods to select tuning parameters and retain more false positives under high dimensionality. This chapter discusses sparse boosting based machine learning methods in the following high-dimensional problems. First, a sparse boosting method to select important biomarkers is studied for the right censored survival data with high-dimensional biomarkers. Then, a two-step sparse boosting method to carry out the variable selection and the model-based prediction is studied for the high-dimensional longitudinal observations measured repeatedly over time. Finally, a multi-step sparse boosting method to identify patient sub-groups that exhibit different treatment effects is studied for the high-dimensional dense longitudinal observations. This chapter intends to solve the problem of how to improve the accuracy and calculation speed of variable selection and parameter estimation in high-dimensional data. It aims to expand the application scope of sparse boosting and develop new methods of high-dimensional survival analysis, longitudinal data analysis, and subgroup analysis, which has great application prospects.

Keywords: sparse boosting, high-dimensional data, machine learning, variable selection, data analysis

1. Introduction

High-dimensional model has become very popular in statistical literature and many new machine learning techniques have been developed to deal with data with very large number of features. In the past decades, researchers have done a great deal of high-dimensional data analysis where the sample size n is relatively small but the number of features p under consideration is extremely large. It is widely known that including irrelevant predictors in the statistical model may result in unstable estimation and dreadful computing issues. Thus, variable selection is crucial to address the challenges. Among all developments, regularization procedures such as LASSO [1], smoothly clipped absolute deviation (SCAD) [2], MCP [3] and their various extensions [4–6] have been thoroughly studied and widely used to perform variable selection and estimation simultaneously in order to improve the prediction accuracy and interpretability of the statistical model. However, those penalized estimation approaches all have some tuning parameters required to be selected by computationally expensive methods like cross-validation.

In recent years, machine learning methods such as boosting have become very prominent for high-dimensional data settings since they can improve the selection accuracy substantially and reduce the chance of including irrelevant features. The original boosting algorithms were proposed by Schapire [7] which is an ensemble method that iteratively combines weaker learners to minimize the expected loss. The major difference among different boosting algorithms is the loss function. For example, AdaBoost [8] has the exponential loss function, L2 boosting [9] has the squared error loss function, sparse boosting [10] has the penalized loss function and HingeBoost [11] has the weighted hinge loss function. Recently, more various versions of boosting algorithms have been proposed. See, for example, Bühlmann and Hothorn [12] for the twin boosting; Komori and Eguchi [13] for the pAUCBoost; Wang [14] for the twin HingeBoost; Zhao [15] for the GSBoosting and Yang and Zou [16] for the ER-Boost. Besides these extensions, much effort has been made in understanding the advantages of boosting such as relatively lower over-fitting risk, smaller computational cost, and simpler adjustment to include additional constraints.

In this chapter we review some sparse boosting based methods for the following high-dimensional problems based on three research papers. First, a sparse boosting method to select important biomarkers is studied for the right censored survival data with high-dimensional biomarkers [17]. Then, a two-step sparse boosting to carry out the variable selection and the model-based prediction is studied for the high-dimensional longitudinal observations measured repeated over time [18]. Finally, a multi-step sparse boosting method to identify patient subgroups that exhibit different treatment effects is studied for the high-dimensional dense longitudinal observations [19]. This chapter intends to solve the problem of how to improve the accuracy and calculation speed of variable selection and parameter estimation in high-dimensional data. It aims to expand the application scope of sparse boosting and develop new methods of high-dimensional survival analysis, longitudinal data analysis, and subgroup analysis, which has great application prospects.

The rest of the chapter is arranged as follows. In Section 2, a sparse boosting method to fit high-dimensional survival data is studied. In Section 3, a two-step sparse boosting approach to carry out variable selection and model-based prediction by fitting high-dimensional models with longitudinal data is studied. In Section 4, a subgroup identification method incorporating multi-step sparse boosting algorithm for high-dimensional dense longitudinal data is studied. Finally, Section 5 provides concluding remarks.

2. Sparse boosting for survival data

Survival time data are usually referred to time-to-event data and they are usually censored. Predicting survival time and identifying the risk factors can be very helpful for patient treatment selection, disease prevention strategy or disease management in evidence-based medicine. A well-known model in survival analysis is the Cox proportional hazards (PH) model [20] which assumes multiplicative covariate effects in the hazards function. Another popular model is the accelerated failure time (AFT) model [21] which assumes that the covariate effect is to accelerate or decelerate the life time of a disease. The coefficients in the regression model have the direct interpretation of the covariate effects on the mean survival time. Recently, researchers developed boosting methods to analyze survival data. For example, Schmid and Hothorn [22] proposed a flexible boosting method for parametric AFT models, and Wang and Wang [23] proposed Buckley-James boosting for survival data with right censoring and high dimensionality.

In this section, a sparse boosting method to fit high-dimensional varying-coefficient AFT models is presented. In particular, the sparse boosting techniques for right censored survival data is studied. In Section 2.1, the varying-coefficient AFT model for survival data is formulated and a detailed sparse boosting algorithm to fit the model is proposed. In Section 2.2, the proposed sparse boosting techniques through simulation studies is evaluated. In Section 2.3, the performance of sparse boosting via a lung cancer data example is examined.

2.1 Methodology

2.1.1 Model and estimation

Let T_i and C_i be the logarithm of survival time and censoring time for the ith subject in a random sample of size n respectively. In reality $Y_i = \min\{T_i, C_i\}$ and the censoring indicator $\delta_i = I(T_i \leq C_i)$ [24] are observed. Denote $\mathbf{X}_i = \left(X_{i,1}, \cdots, X_{i,p-1}\right)$ to be the corresponding (p-1)-dimensional predictors such as gene expressions or biomarkers for the ith subject and U_i to be the univariate index variable. Our observed data set $\left\{(\mathbf{X}_i, \delta_i, Y_i, U_i) : \mathbf{X}_i \in \mathbb{R}^{p-1}, \delta_i \in \{0,1\}, Y_i \in \mathbb{R}, U_i \in \mathbb{R}, i = 1, 2, \cdots, n\right\}$ is an independently and identically distributed random sample from $(\mathbf{X}, \delta, Y, U)$. The varying-coefficient AFT model is:

$$T_i = \beta_0(U_i) + \sum_{j=1}^{p-1} X_{i,j}\beta_j(U_i) + \varepsilon_i, \quad i = 1, \dots, n, \tag{1}$$

where $\beta_0(.), \beta_1(.), \cdots, \beta_{p-1}(.)$ are the unknown varying-coefficient functions of confounder U and ε_i is the random error with $E(\varepsilon_i | \mathbf{X}_i, U_i) = 0$.

A weighted least squares estimation approach is adopted. Let w_i's be the Kaplan–Meier weights [25], which are the jumps in the Kaplan–Meier estimator computed as $w_1 = \frac{\delta_{(1)}}{n}$ and $w_i = \frac{\delta_{(i)}}{n-i+1}\prod_{j=1}^{i-1}\left(\frac{n-j}{n-j+1}\right)^{\delta_{(j)}}, i = 2, \dots, n$. Let $Y_{(1)} \leq \cdots \leq Y_{(n)}$ be the order statistics of Y_i's, $\delta_{(1)}, \cdots, \delta_{(n)}$ be the corresponding censoring indicators of the ordered Y_i's, and $X_{(1),j}, \cdots, X_{(n),j}, j = 1, \cdots, p-1$ and $U_{(1)}, \cdots, U_{(n)}$ are defined similarly. Then the weighed least squares loss function is

$$\sum_{i=1}^{n} w_i \left(Y_{(i)} - \beta_0\left(U_{(i)}\right) - \sum_{j=1}^{p-1} X_{(i),j}\beta_j\left(U_{(i)}\right) \right)^2. \tag{2}$$

Let $B(.) = (B_1(.), \dots, B_L(.))^T$ be an equal-spaced B-spline basis, where L is the dimension of the basis. Under certain smoothness conditions, the Curry-Schonberg theorem [26] implies that for every smooth function $\beta_j(.)$, it can be approximated by

$$\beta_j(.) \approx B^T(.)\gamma_j, \quad j = 0, \cdots, p-1, \tag{3}$$

where γ_j is a vector of length L. Then the weighted least squares loss function Eq. (2) can be approximated by

$$\sum_{i=1}^{n} w_i \left(Y_{(i)} - B^T(U_{(i)})\gamma_0 - \sum_{j=1}^{p-1} X_{(i),j} B^T(U_{(i)})\gamma_j \right)^2. \tag{4}$$

Denote by $\tilde{Y} = \left(Y_{(1)}, \cdots, Y_{(n)}\right)^T$, $X_{(i),0} = 1$ for $i = 1, \cdots, n$, $\tilde{\mathbf{X}}_j = \left(B(U_{(1)})X_{(1),j}, \cdots, B(U_{(n)})X_{(n),j}\right)^T$, $\tilde{\mathbf{X}} = (\tilde{\mathbf{X}}_0, \cdots, \tilde{\mathbf{X}}_{p-1})$, $W = \text{diag}(w_1, \cdots, w_n)$ and $\gamma = \left(\gamma_0^T, \cdots, \gamma_{p-1}^T\right)^T$. Then the objective function Eq. (4) may be written in the following matrix form:

$$(\tilde{Y} - \tilde{\mathbf{X}}\gamma)^T W (\tilde{Y} - \tilde{\mathbf{X}}\gamma). \tag{5}$$

The estimation may yield close-form solution for the coefficients when dimensionality p is small or moderate. With high dimensionality the solution cannot be easily achieved. Let $\gamma^{[\hat{K}]} = \left(\left(\gamma_0^{[\hat{K}]}\right)^T, \cdots, \left(\gamma_{p-1}^{[\hat{K}]}\right)^T\right)^T$ be the estimator of γ from sparse boosting approach with weighted square loss function Eq. (5), and $\hat{K}$ is the estimated number of stopping iterations. Then the estimates of coefficient function are given by

$$\hat{\beta}_j(u) = B^T(u)\gamma_j^{[\hat{K}]}, \quad j = 0, \cdots, p-1. \tag{6}$$

Instead of using the regularized estimation approaches, a sparse boosting method to estimate $\gamma^{[\hat{K}]}$ is presented in the following subsection.

2.1.2 Sparse boosting techniques

The key idea of sparse boosting is to replace the empirical risk function in L2 boosting with the penalized empirical risk function which is a combination of squared loss and the trace of boosting operator as a measure of boosting complexity, and then perform gradient descent in a function space iteratively. Thus sparse boosting produces sparser models compared to L2 boosting. The g-prior minimum description length (gMDL) proposed by [27] can be used as the penalized empirical risk function to estimate the update criterion in each iteration and the stopping criterion. The gMDL takes the form:

$$\text{gMDL}(RSS, \text{trace}(\mathcal{B})) = \log(S) + \frac{\text{trace}(\mathcal{B})}{n} \log\left(\frac{\tilde{Y}^T\tilde{Y} - RSS}{\text{trace}(\mathcal{B}) \times S}\right),$$
$$S = \frac{RSS}{n - \text{trace}(\mathcal{B})}. \tag{7}$$

Here RSS is the residual sum of squares and $\mathcal{B}$ is the boosting operator. The model that achieves the shortest description of data will be selected. The advantage is that it has a data-dependent penalty for each dimension since it is explicitly given as a function of data only, thus the selection of the tuning parameter can be avoided.

The sparse boosting procedure is described in details. The initial value of γ is set to be a zero vector, i.e. $\gamma^{[k]} = \mathbf{0}$ for $k = 0$, while in each of the kth iteration ($1 \le k \le K$ for K being the total number of iterations) only the current residual $R^{[k]} = \tilde{Y} - \tilde{\mathbf{X}}\gamma^{[k-1]}$ is used to regress every jth working element $\tilde{\mathbf{X}}_j, j = 0, \cdots, p-1$. The fit denoted by $\hat{\lambda}_j^{[k]}$ can be obtained by minimizing the weighted squared loss function $(R^{[k]} - \tilde{\mathbf{X}}_j\lambda)^T W(R^{[k]} - \tilde{\mathbf{X}}_j\lambda)$ with respect to λ. Hence the weighted least squared estimate is $\hat{\lambda}_j^{[k]} = \left[(\tilde{\mathbf{X}}_j)^T W(\tilde{\mathbf{X}}_j)\right]^{-1}(\tilde{\mathbf{X}}_j)^T WR^{[k]}$, the corresponding hat matrix is $\mathcal{H}_j = (\tilde{\mathbf{X}}_j)\left[(\tilde{\mathbf{X}}_j)^T W(\tilde{\mathbf{X}}_j)\right]^{-1}(\tilde{\mathbf{X}}_j)^T W$ and the weighted residual sum of squares is $RSS_j^{[k]} = \left(R^{[k]} - \tilde{\mathbf{X}}_j\hat{\lambda}_j^{[k]}\right)^T W\left(R^{[k]} - \tilde{\mathbf{X}}_j\hat{\lambda}_j^{[k]}\right)$. The selected component $\hat{s}_k$ can be obtained by:

$$\hat{s}_k = \operatorname{argmin}_{0 \leq j \leq p-1} \text{gMDL}\left(RSS_j^{[k]}, \text{trace}\left(\mathcal{B}_j^{[k]}\right)\right), \tag{8}$$

where $\mathcal{B}_j^{[1]} = \mathcal{H}_j$ and $\mathcal{B}_j^{[k]} = I - (I - \mathcal{H}_j)(I - \nu\mathcal{H}_{\hat{s}_{k-1}})\cdot\cdots\cdot(I - \nu\mathcal{H}_{\hat{s}_1})$ for $k > 1$ is the boosting operator for selecting jth component in the kth iteration. Therefore, at each iteration there is only one working component $\tilde{\mathbf{X}}_{\hat{s}_k}$ to be chosen, and only the corresponding coefficient vector $\gamma_{\hat{s}_k}^{[k]}$ changes, i.e. $\gamma_{\hat{s}_k}^{[k]} = \gamma_{\hat{s}_k}^{[k-1]} + \nu\hat{\lambda}_{\hat{s}_k}^{[k]}$, where ν is the step size, while all the other $\gamma_j^{[k]}$ for $j \neq \hat{s}_k$ remain the same. This process is repeated for K iterations and estimate the stopping iteration K by.

$$\hat{K} = \operatorname{argmin}_{1 \leq k \leq K} \text{gMDL}\left(RSS_{\hat{s}_k}^{[k]}, \text{trace}\left(\mathcal{B}^{[k]}\right)\right), \tag{9}$$

where $\mathcal{B}^{[k]} = I - (I - \nu\mathcal{H}_{\hat{s}_k})\cdot\cdots\cdot(I - \nu\mathcal{H}_{\hat{s}_1})$.

From this sparse boosting procedure, the estimator of γ is obtained as $\gamma^{[\hat{K}]} = \left(\left(\gamma_0^{[\hat{K}]}\right)^T, \cdots, \left(\gamma_{p-1}^{[\hat{K}]}\right)^T\right)^T$. The sparse boosting algorithm for the varying-coefficient AFT model can be summarized as follows:

Sparse Boosting Algorithm for Varying-Coefficient AFT Model.

a. Initialization. Set $k = 0$ and $\gamma_0^{[k]} = \mathbf{0}, \cdots, \gamma_{p-1}^{[k]} = \mathbf{0}$ (component-wise).

b. Iteration. $k = k + 1$. Compute $\hat{s}_k = \operatorname{argmin}_{0 \leq j \leq p-1} \text{gMDL}\left(RSS_j^{[k]}, \text{trace}\left(\mathcal{B}_j^{[k]}\right)\right)$, where $\mathcal{B}_j^{[1]} = \mathcal{H}_j$ and $\mathcal{B}_j^{[k]} = I - (I - \mathcal{H}_j)(I - \nu\mathcal{H}_{\hat{s}_{k-1}})\cdot\cdots\cdot(I - \nu\mathcal{H}_{\hat{s}_1})$ for $k > 1$.

c. Update. $\gamma_{\hat{s}_k}^{[k]} = \gamma_{\hat{s}_k}^{[k-1]}$ for $j \neq \hat{s}_k$ and $\gamma_{\hat{s}_k}^{[k]} = \gamma_{\hat{s}_k}^{[k-1]} + \nu\hat{\lambda}_{\hat{s}_k}^{[k]}$, where ν is the step size.

d. Iteration. Repeat step (b)-(c) for K iterations.

e. Stopping. Estimate $\hat{K} = \operatorname{argmin}_{1 \leq k \leq K} \text{gMDL}\left(RSS_{\hat{s}_k}^{[k]}, \text{trace}\left(\mathcal{B}^{[k]}\right)\right)$, where $\mathcal{B}^{[k]} = I - (I - \nu\mathcal{H}_{\hat{s}_k})\cdot\cdots\cdot(I - \nu\mathcal{H}_{\hat{s}_1})$.Thus, $\gamma^{[\hat{K}]} = \left(\left(\gamma_0^{[\hat{K}]}\right)^T, \cdots, \left(\gamma_{p-1}^{[\hat{K}]}\right)^T\right)^T$ is the estimate for γ and $\hat{\beta}_j(u) = B^T(u)\gamma_j^{[\hat{K}]}, j = 0, \cdots, p-1$ are the estimators for varying coefficients. The final estimator of $\tilde{Y}$ is $\hat{\tilde{Y}}^{[\hat{K}]} = \tilde{\mathbf{X}}\gamma^{[\hat{K}]}$.

According to [10] and references therein, the selection of step size ν is of minor importance as long as it is small. A smaller value of ν achieves higher prediction accuracy while requires a larger number of boosting iterations and more computing time. A typical value used in literature is $\nu = 0.1$.

2.2 Simulation

The performance of the above sparse boosting algorithm is evaluated by studying their performance on simulated data. L2 boosting and sparse boosting methods are compared in their performance of variable selection and function estimation. Sparse boosting method is what we present in this section while L2 boosting method is a relatively simpler version and may not achieve sparse solution in general.

The simulation results from [17] show that both boosting methods can identify important variables while sparse boosting selects much fewer irrelevant variables than L2 boosting. Although in-sample prediction errors (defined as $\sum_{i=1}^{n}\delta_i\left(Y_i - Y_i^{[\hat{K}]}\right)^2/\sum_{i=1}^{n}\delta_i)$ using L2 boosting is a little bit smaller than using sparse boosting since the former has larger model sizes, the average of root mean integrated squared errors (defined as $\sqrt{\frac{1}{n}\sum_{j=0}^{5}\sum_{i=1}^{n}\left(\beta_j(u_i) - \hat{\beta}_j(u_i)\right)^2}$) using sparse boosting is much smaller than that using L2 boosting. Furthermore, when the smoothness assumption in Curry-Schonberg theorem is violated for the coefficient functions, the performance of variable selection remains good. In summary, sparse boosting outperforms L2 boosting in terms of parameter estimation and variable selection.

2.3 Lung cancer data analysis

Lung cancer is the top cancer killer for people in the U.S. Identifying relevant gene expressions in lung cancer is important for treatment and prevention. Our data is from a large multi-site blinded validation study [28] with 442 lung adenocarcinomas. Age is treated as the potential confounder in this analysis, since it is usually strongly correlated with survival time [29]. After removing missing measurements and predictors in overall survival, a total of 439 patients are left in the analysis. For each patient, 22,283 gene expressions are available. The median follow-up time is 46 months (range: 0.03 to 204 months) with the overall censoring rate 46.47 %. The median age at diagnosis is 65 years (range: 33 to 87 years). After adopting a marginal screening procedure to screen out irrelevant genes, variable selection approaches are used to identify important genes associated with lung cancer. With the aim of comparison, except L2 boosting and the proposed sparse boosting, the following existing variable selection approaches for constant-coefficient AFT models are also considered: Buckley-James boosting with linear least squares [23], Buckley-James twin boosting with linear least squares [23], Buckley-James regression with elastic net penalty [30] and SCAD penalty respectively.

The results from [17] show that L2 boosting and sparse boosting for varying-coefficient AFT model not only produce relatively sparser model, but also have smaller in-sample and out-of-sample prediction error compared to the four methods for constant-coefficient AFT model. Again, sparse boosting produce even sparser model than L2 boosting. In conclusion, including age in the varying-coefficient AFT model could lead to more accurate estimate than constant-coefficient AFT model and the proposed sparse boosting method for varying-coefficient AFT model has good performance in terms of estimation, prediction as well as sparsity.

3. Two-step sparse boosting for longitudinal data

Longitudinal data contain repeated measurements collected from the same respondents over time. The assumption that all measurements are independent does not hold for such data. One important question in longitudinal analysis is how to make efficient inference by taking into account of the within subjects correlation. This question has been investigated in depth by many researchers [31, 32] for parametric models. Semiparametric and nonparametric models for longitudinal data are also presented in the literature, see [33, 34]. Recently, there are some development on longitudinal data with high-dimensionalilty using varying-coefficient models [35, 36]. All previous studies adopted the penalty methods.

In this section, a two-step sparse boosting approach is presented to preform the variable selection and the model-based prediction. Specifically, high-dimensional varying-coefficient models with longitudinal data will be considered. In the first step, the sparse boosting approach is utilized to obtain an estimate of the correlation structure. In the second step, the within-subject correlation structure is considered and variable selection and coefficients estimation are achieved by sparse boosting again. The rest of this section is arranged as follows. In Section 3.1, the varying-coefficient model for longitudinal data is formulated and a two-step sparse boosting algorithm is presented. In Section 3.2, simulation studies are conducted to illustrate the validity of the two-step sparse boosting method. In Section 3.3, the performance of two-stage method is assessed by studying yeast cell cycle gene expression data.

3.1 Methodology

3.1.1 Model and estimation

Let Y_{ij} be the continuous outcome for the jth measurement of individual i taken at time $t_{ij} \in T$, where T is the time interval on which the measurements are taken. Denote $\mathbf{X}_{ij} = \left(X_{ij,1}, \cdots, X_{ij,p-1}\right)$ to be the corresponding (p-1)-dimensional covariate vector. The varying-coefficient model which can capture the dynamical impacts of the covariates on the response variable is considered:

$$Y_{ij} = \beta_0\left(t_{ij}\right) + \sum_{d=1}^{p-1} X_{ij,d}\beta_d\left(t_{ij}\right) + \varepsilon_{ij}, \quad i = 1, \cdots, n, j = 1, \cdots, n_i, \tag{10}$$

where $\beta_0(.), \beta_1(.), \cdots, \beta_{p-1}(.)$ are the unknown smooth coefficient functions of time and $\varepsilon_i = \left(\varepsilon_{i1}, \cdots, \varepsilon_{in_i}\right)^T, i = 1, \cdots, n$ are multivariate error terms with mean zero. Errors are assumed to be uncorrelated for different i, but components of ε_i are correlated with each other. Without loss of generality, the balanced longitudinal study is considered in the following implementation, i.e., $t_{ij} = t_{kj}$, and $n_i = m$ for all i.

The estimation procedure is presented below. In the first step, the within-subject correlation is ignored first and the coefficients are estimated by minimizing the following least squares loss function:

$$\sum_{i=1}^{n}\sum_{j=1}^{m}\left(Y_{ij} - \beta_0\left(t_{ij}\right) - \sum_{d=1}^{p-1} X_{ij,d}\beta_d\left(t_{ij}\right)\right)^2. \tag{11}$$

The B-spline basis is used to estimate the coefficient functions $\beta_0(.), \beta_1(.), \cdots, \beta_{p-1}(.)$. Denote $B(.) = (B_1(.), \ldots, B_L(.))^T$ to be an equal-spaced

B-spline basis of dimension L. Under certain smoothness assumptions, function $\beta_d(.)$ can be approximated by

$$\beta_d(.) \approx B^T(.)\gamma_d, \quad d = 0, \cdots, p-1, \tag{12}$$

where γ_d is a loading vector of length L. Then the least squares loss function Eq. (11) is close to

$$\sum_{i=1}^{n}\sum_{j=1}^{m}\left(Y_{ij} - B^T(t_{ij})\gamma_0 - \sum_{d=1}^{p-1} X_{ij,d}B^T(t_{ij})\gamma_d\right)^2. \tag{13}$$

Further denote $Y_i = (Y_{i1}, \cdots, Y_{im})^T$, $Y = \left(Y_1^T, \cdots, Y_n^T\right)^T$, $X_{ij,0} = 1$, $\tilde{\mathbf{X}}_{i,d} = (B(t_{i1})X_{i1,d}, \cdots, B(t_{im})X_{im,d})^T$, $\tilde{\mathbf{X}}_i = (\tilde{\mathbf{X}}_{i,0}, \cdots, \tilde{\mathbf{X}}_{i,p-1})$, $\tilde{\mathbf{X}} = \left(\tilde{\mathbf{X}}_1^T, \cdots, \tilde{\mathbf{X}}_n^T\right)^T$ and $\gamma = \left(\gamma_0^T, \cdots, \gamma_{p-1}^T\right)^T$. Then the target function Eq. (13) can be expressed in the matrix format:

$$\sum_{i=1}^{n}(Y_i - \tilde{\mathbf{X}}_i\gamma)^T(Y_i - \tilde{\mathbf{X}}_i\gamma) \equiv (Y - \tilde{\mathbf{X}}\gamma)^T(Y - \tilde{\mathbf{X}}\gamma). \tag{14}$$

Denote $\gamma^{\left[\widehat{K_1}\right]}$ to be the estimator of γ by sparse boosting with squared loss function Eq. (14) being loss function, where $\widehat{K_1}$ is the estimated stopping iterations in this step. There is no exact closed form for $\gamma^{\left[\widehat{K_1}\right]}$ since it is derived from an iterative algorithm. However it can be evaluated very fast in a computer implementation. The detailed algorithm will be presented in the next subsection.

The first step coefficient estimates are given by

$$\tilde{\beta}_d(t) = B^T(t)\gamma_d^{\left[\widehat{K_1}\right]}, \quad d = 0, \cdots, p-1. \tag{15}$$

Write $\hat{\varepsilon}_i = Y_i - \tilde{\mathbf{X}}_i\gamma^{\left[\widehat{K_1}\right]}$, $i = 1, \cdots, n$. The $m \times m$ covariance matrix $Cov(Y_i) \equiv \Sigma$ can be estimated by the following empirical estimator

$$\widehat{\sum} = \frac{1}{n}\sum_{i=1}^{n}\hat{\varepsilon}_i\hat{\varepsilon}_i^T. \tag{16}$$

In the second step, the estimated correlation structure within repeated measurements is taken into account to form the weighted least squares loss function as follows:

$$\sum_{i=1}^{n}(Y_i - \tilde{\mathbf{X}}_i\gamma^\star)^T\hat{\Sigma}^{-1}(Y_i - \tilde{\mathbf{X}}_i\gamma^\star) \equiv (Y - \tilde{\mathbf{X}}\gamma^\star)^T W(Y - \tilde{\mathbf{X}}\gamma^\star), \tag{17}$$

where $W = \text{diag}\left(\hat{\Sigma}^{-1}, \cdots, \hat{\Sigma}^{-1}\right)$ is the estimated $(n \times m) \times (n \times m)$ weight matrix. Denote $\gamma^{\star\left[\widehat{K_2}\right]}$ to be the estimator of $\gamma^\star$ by sparse boosting with weighted loss function Eq. (17) being the loss function, where $\widehat{K_2}$ is the estimated stopping iterations in the second step. Then the coefficient estimates from the second step are given by

$$\hat{\beta}_d(t) = B^T(t)\gamma_d^{\star\left[\widehat{K_2}\right]}, \quad d = 0, \cdots, p-1. \tag{18}$$

The reliable estimates for the coefficient functions could then be obtained. More details about how to use sparse boosting to get $\gamma^{[\widehat{K_1}]}$ and $\gamma^{\star [\widehat{K_2}]}$ are provided in the following subsection.

3.1.2 Two-step sparse boosting techniques

gMDL can be adopted as the penalized empirical risk function to estimate the update criterion in each iteration and the stopping criterion. gMDL can be expressed in the following form:

$$\text{gMDL}(RSS, \text{trace}(\mathcal{B})) = \log(F) + \frac{\text{trace}(\mathcal{B})}{n \times m} \log\left(\frac{Y^T Y - RSS}{\text{trace}(\mathcal{B}) \times F}\right), \qquad F = \frac{RSS}{n \times m - \text{trace}(\mathcal{B})}, \tag{19}$$

where $\mathcal{B}$ is the boosting operator and RSS is the residual sum of squares.

The two-step sparse boosting approach is presented more specifically. In the first step, the start value of γ is set to zero vector, i.e. $\gamma^{[0]} = \mathbf{0}$, and in each of the k_1th iteration ($0 < k_1 \le K_1$, and K_1 is the maximum number of iterations considered in the first step), the residual $R^{[k_1]} = Y - \tilde{\mathbf{X}}\gamma^{[k_1 - 1]}$ in present iteration is used to fit each of the dth component $\tilde{\mathbf{X}}_{,d} = \left(\tilde{\mathbf{X}}_{1,d}^T, \cdots, \tilde{\mathbf{X}}_{n,d}^T\right)^T, d = 0, \cdots, p-1$ by treating all the within-subject observations uncorrelated. Then the fit denoted by $\hat{\lambda}_d^{[k_1]}$ can be calculated by minimizing the squared loss function $(R^{[k_1]} - \tilde{\mathbf{X}}_{,d}\lambda)^T (R^{[k_1]} - \tilde{\mathbf{X}}_{,d}\lambda)$ with respect to λ. Therefore, the least squares estimate is $\hat{\lambda}_d^{[k_1]} = \left[(\tilde{\mathbf{X}}_{,d})^T (\tilde{\mathbf{X}}_{,d})\right]^{-1} (\tilde{\mathbf{X}}_{,d})^T R^{[k_1]}$, the corresponding hat matrix is $\mathcal{H}_d = (\tilde{\mathbf{X}}_{,d}) \left[(\tilde{\mathbf{X}}_{,d})^T (\tilde{\mathbf{X}}_{,d})\right]^{-1} (\tilde{\mathbf{X}}_{,d})^T$ and the residual sum of squares is $RSS_d^{[k_1]} = \left(R^{[k_1]} - \tilde{\mathbf{X}}_{,d}\hat{\lambda}_d^{[k_1]}\right)^T \left(R^{[k_1]} - \tilde{\mathbf{X}}_{,d}\hat{\lambda}_d^{[k_1]}\right)$. The chosen element $\hat{s}_{k_1}$ is attained by:

$$\hat{s}_{k_1} = \text{argmin}_{0 \le d \le p-1} \text{gMDL}\left(RSS_d^{[k_1]}, \text{trace}\left(\mathcal{B}_d^{[k_1]}\right)\right), \tag{20}$$

where $\mathcal{B}_d^{[1]} = \mathcal{H}_d$ and $\mathcal{B}_d^{[k_1]} = I - (I - \mathcal{H}_d)\left(I - \nu\mathcal{H}_{\hat{s}_{k_1 - 1}}\right) \cdot \cdots \cdot (I - \nu\mathcal{H}_{\hat{s}_1})$ for $k_1 > 1$ is the first step boosting operator for choosing dth element in the k_1th iteration. Hence, there is an unique element $\tilde{\mathbf{X}}_{,\hat{s}_{k_1}}$ to be selected at each iteration, and only the corresponding coefficient vector $\gamma_{\hat{s}_{k_1}}^{[k_1]}$ changes, i.e., $\gamma_{\hat{s}_{k_1}}^{[k_1]} = \gamma_{\hat{s}_{k_1}}^{[k_1 - 1]} + \nu\hat{\lambda}_{\hat{s}_{k_1}}^{[k_1]}$, where ν is the pre-specified step-size parameter. All the other $\gamma_d^{[k_1]}$ for $d \ne \hat{s}_{k_1}$ keep unchanged. This procedure is repeated for K_1 times and the number of iterations K_1 can be estimated by

$$\widehat{K_1} = \text{argmin}_{1 \le k_1 \le K_1} \text{gMDL}\left(RSS_{\hat{s}_{k_1}}^{[k_1]}, \text{trace}\left(\mathcal{B}^{[k_1]}\right)\right), \tag{21}$$

where $\mathcal{B}^{[k_1]} = I - \left(I - \nu\mathcal{H}_{\hat{s}_{k_1}}\right) \cdot \cdots \cdot (I - \nu\mathcal{H}_{\hat{s}_1})$.

From the first step of sparse boosting, the estimator of γ is obtained by $\gamma^{[\widehat{K_1}]} = \left(\left(\gamma_0^{[\widehat{K_1}]}\right)^T, \cdots, \left(\gamma_{p-1}^{[\widehat{K_1}]}\right)^T\right)^T$. Then the weight matrix W can be easily obtained too.

In the second step, sparse boosting is used again by taking into account of the correlation structure estimator for the repeated measurements estimated in the first step. The initial value of $\gamma^\star$ is set to be the coefficient estimator from the first step of sparse boosting, i.e. $\gamma^{\star[0]} = \gamma^{[\widehat{K_1}]}$, and in each of the k_2th iteration ($0 < k_2 \le K_2$, and K_2 is the maximum number of iterations under consideration in the second step), the residual $R^{\star[k_2]} = Y - \tilde{\mathbf{X}}\gamma^{\star[k_2-1]}$ in current iteration is used to fit each of the dth working element $\tilde{\mathbf{X}}_{,d}, d = 0, \cdots, p-1$ by incorporating the within-subject correlation estimator from the first step. Then the fit denoted by $\hat{\lambda}_d^{\star[k_2]}$ can be obtained by minimizing the weighted squared loss function $\left(R^{\star[k_2]} - \tilde{\mathbf{X}}_{,d}\lambda\right)^T W\left(R^{\star[k_2]} - \tilde{\mathbf{X}}_{,d}\lambda\right)$ with respect to λ. Thus, the weighted least squares estimate is $\hat{\lambda}_d^{\star[k_2]} = \left[\left(\tilde{\mathbf{X}}_{,d}\right)^T W\left(\tilde{\mathbf{X}}_{,d}\right)\right]^{-1}\left(\tilde{\mathbf{X}}_{,d}\right)^T W R^{\star[k_2]}$, the corresponding hat matrix is $\mathcal{H}_d^\star = \left(\tilde{\mathbf{X}}_{,d}\right)\left[\left(\tilde{\mathbf{X}}_{,d}\right)^T W\left(\tilde{\mathbf{X}}_{,d}\right)\right]^{-1}\left(\tilde{\mathbf{X}}_{,d}\right)^T W$ and the weighted residual sum of squares is $RSS_d^{\star[k_2]} = \left(R^{\star[k_2]} - \tilde{\mathbf{X}}_{,d}\hat{\lambda}_d^{\star[k_2]}\right)^T W\left(R^{\star[k_2]} - \tilde{\mathbf{X}}_{,d}\hat{\lambda}_d^{\star[k_2]}\right)$. The chosen element $\hat{s}_{k_2}$ can be obtained by:

$$\hat{s}_{k_2} = \operatorname{argmin}_{0 \le d \le p-1} \text{gMDL}\left(RSS_d^{\star[k_2]}, \text{trace}\left(\mathcal{B}_d^{\star[k_2]}\right)\right), \tag{22}$$

where $\mathcal{B}_d^{\star[1]} = I - \left(I - \mathcal{B}^{[\widehat{K_1}]}\right)\left(I - \mathcal{H}_d^\star\right)$ and $\mathcal{B}_d^{\star[k_2]} = I - \left(I - \mathcal{B}^{[\widehat{K_1}]}\right)\left(I - \mathcal{H}_d^\star\right)\left(I - \nu\mathcal{H}_{\hat{s}_{k_2-1}}^\star\right)\cdots\left(I - \nu\mathcal{H}_{\hat{s}_1}^\star\right)$ for $k_2 > 1$ is the second step boosting operator for choosing dth element in the k_2th iteration. Thus, there is an unique element $\tilde{\mathbf{X}}_{,\hat{s}_{k_2}}$ to be selected at each time, and only the corresponding coefficient vector $\gamma_{\hat{s}_{k_2}}^{\star[k_2]}$ change, i.e., $\gamma_{\hat{s}_{k_2}}^{\star[k_2]} = \gamma_{\hat{s}_{k_2}}^{\star[k_2-1]} + \nu\hat{\lambda}_{\hat{s}_{k_2}}^{\star[k_2]}$. While all the other $\gamma_d^{\star[k_2]}$ for $d \ne \hat{s}_{k_2}$ remain the same. This procedure is repeated for K_2 times and the estimated stopping iterations $\widehat{K_2}$ is

$$\widehat{K_2} = \operatorname{argmin}_{1 \le k_2 \le K_2} \text{gMDL}\left(RSS_{\hat{s}_{k_2}}^{\star[k_2]}, \text{trace}\left(\mathcal{B}^{\star[k_2]}\right)\right), \tag{23}$$

where $\mathcal{B}^{\star[k_2]} = I - \left(I - \mathcal{B}^{[\widehat{K_1}]}\right)\left(I - \nu\mathcal{H}_{\hat{s}_{k_2}}^\star\right)\cdots\left(I - \nu\mathcal{H}_{\hat{s}_1}^\star\right)$.

From the second step of sparse boosting, the estimator of $\gamma^\star$ is arrived by $\gamma^{\star[\widehat{K_2}]} = \left(\left(\gamma_0^{\star[\widehat{K_2}]}\right)^T, \cdots, \left(\gamma_{p-1}^{\star[\widehat{K_2}]}\right)^T\right)^T$. The two-step sparse boosting algorithm for varying-coefficient model with longitudinal data can be summarized in the following form:

Two-step Sparse Boosting Algorithm with Longitudinal Data.

Step I: Use sparse boosting to estimate covariance matrix.

a. Initialization. Let $k_1 = 0$ and $\gamma_0^{[k_1]} = \mathbf{0}, \cdots, \gamma_{p-1}^{[k_1]} = \mathbf{0}$.

b. Increase k_1 by 1. Calculate $\hat{s}_{k_1} = \operatorname{argmin}_{0 \le d \le p-1} \text{gMDL}\left(RSS_d^{[k_1]}, \text{trace}\left(\mathcal{B}_d^{[k_1]}\right)\right)$, where $\mathcal{B}_d^{[1]} = \mathcal{H}_d$ and $\mathcal{B}_d^{[k_1]} = I - (I - \mathcal{H}_d)\left(I - \nu\mathcal{H}_{\hat{s}_{k_1-1}}\right)\cdots(I - \nu\mathcal{H}_{\hat{s}_1})$ for $k_1 > 1$.

c. Update. $\gamma_{\hat{s}_{k_1}}^{[k_1]} = \gamma_{\hat{s}_{k_1}}^{[k_1-1]}$ for $d \neq \hat{s}_{k_1}$ and $\gamma_{\hat{s}_{k_1}}^{[k_1]} = \gamma_{\hat{s}_{k_1}}^{[k_1-1]} + \nu\hat{\lambda}_{\hat{s}_{k_1}}^{[k_1]}$, where ν is the step-size parameter.

d. Iteration. Repeat step (b)-(c) for some large iteration number K_1.

e. Stopping. The optimal iteration number can be taken as $\widehat{K_1} = \text{argmin}_{1 \le k_1 \le K_1} \text{gMDL}\left(RSS_{\hat{s}_{k_1}}^{[k_1]}, \text{trace}\left(\mathcal{B}^{[k_1]}\right)\right)$, where $\mathcal{B}^{[k_1]} = I - \left(I - \nu\mathcal{H}_{\hat{s}_{k_1}}\right) \cdot \cdots \cdot \left(I - \nu\mathcal{H}_{\hat{s}_1}\right)$.

Thus, $\gamma^{\left[\widehat{K_1}\right]} = \left(\left(\gamma_0^{\left[\widehat{K_1}\right]}\right)^T, \cdots, \left(\gamma_{p-1}^{\left[\widehat{K_1}\right]}\right)^T\right)^T$ is the first step estimator for γ from sparse boosting and $\tilde{\beta}_d(t) = B^T(t)\gamma_d^{\left[\widehat{K_1}\right]}$, $d = 0, \cdots, p-1$ are the varying coefficient estimates ignoring the within-subject correlation. $Cov(Y_i)$ can be estimated by

$$\hat{\Sigma} = \frac{1}{n}\sum_{i=1}^{n}\left(Y_i - \tilde{\mathbf{X}}_i\gamma^{\left[\widehat{K_1}\right]}\right)\left(Y_i - \tilde{\mathbf{X}}_i\gamma^{\left[\widehat{K_1}\right]}\right)^T.$$

Step II: Use sparse boosting again by incorporating covariance matrix estimator.

a. Initialization. Let $k_2 = 0$ and $\gamma^{\star[k_2]} = \gamma^{\left[\widehat{K_1}\right]}$.

b. Increase k_2 by 1. Calculate $\hat{s}_{k_2} = \text{argmin}_{0 \le d \le p-1} \text{g MDL}\left(RSS_d^{\star[k_2]}, \text{trace}\left(\mathcal{B}_d^{\star[k_2]}\right)\right)$, where $\mathcal{B}_d^{\star[1]} = I - \left(I - \mathcal{B}^{\left[\widehat{K_1}\right]}\right)\left(I - \mathcal{H}_d^{\star}\right)$ and $\mathcal{B}_d^{\star[k_2]} = I - \left(I - \mathcal{B}^{\left[\widehat{K_1}\right]}\right)\left(I - \mathcal{H}_d^{\star}\right)\left(I - \nu\mathcal{H}_{\hat{s}_{k_2-1}}^{\star}\right) \cdot \cdots \cdot \left(I - \nu\mathcal{H}_{\hat{s}_1}^{\star}\right)$ for $k_2 > 1$.

c. Update. $\gamma_{\hat{s}_{k_2}}^{\star[k_2]} = \gamma_{\hat{s}_{k_2}}^{\star[k_2-1]}$ for $d \neq \hat{s}_{k_2}$ and $\gamma_{\hat{s}_{k_2}}^{\star[k_2]} = \gamma_{\hat{s}_{k_2}}^{\star[k_2-1]} + \nu\hat{\lambda}_{\hat{s}_{k_2}}^{\star[k_2]}$.

d. Iteration. Repeat step (b)-(c) for some large iteration number K_2.

e. Stopping. The optimal iteration number can be taken as $\widehat{K_2} = \text{argmin}_{1 \le k_2 \le K_2} \text{gMDL}\left(RSS_{\hat{s}_{k_2}}^{\star[k_2]}, \text{trace}\left(\mathcal{B}^{\star[k_2]}\right)\right)$, where $\mathcal{B}^{\star[k_2]} = I - \left(I - \mathcal{B}^{\left[\widehat{K_1}\right]}\right)\left(I - \nu\mathcal{H}_{\hat{s}_{k_2}}^{\star}\right) \cdot \cdots \cdot \left(I - \nu\mathcal{H}_{\hat{s}_1}^{\star}\right)$.

Therefore, $\gamma^{\star\left[\widehat{K_2}\right]} = \left(\left(\gamma_0^{\star\left[\widehat{K_2}\right]}\right)^T, \cdots, \left(\gamma_{p-1}^{\star\left[\widehat{K_2}\right]}\right)^T\right)^T$ and $\hat{\beta}_d(t) = B^T(t)\gamma_d^{\star\left[\widehat{K_2}\right]}$, $d = 0, \cdots, p-1$ are the final estimator for $\gamma^\star$ and varying coefficient estimates by the two-step sparse boosting. The final estimate for Y is $\hat{Y} = \tilde{\mathbf{X}}\gamma^{\star\left[\widehat{K_2}\right]}$.

3.2 Simulation

Simulation studies are conducted to evaluate the performance of the above two-step sparse boosting algorithm. The following four methods are compared in terms of variable selection and function estimation performance. M1: two-step L2 boosting (use squared loss for update criterion and gMDL for stopping criterion); M2: two-step sparse boosting; M3: two-step lasso (performs lasso regression in the first step to calculate the estimated within-subject correlation structure using Eq. (14), and use lasso regression in the second step by taking into account of the estimated correlation structure) and M4: two-step elastic net regression (similar as M3 with the elastic net mixing parameter 0.5).

The simulation results from [18] show that all methods are able to identify important variables. However, in terms of sparsity, the two-step sparse boosting method preforms best with smallest number of false positives. Both penalization methods select much more irrelevant variables than boosting methods, with elastic net selects the most. For two-step sparse boosting, results of variable selection are quite stable from step I to step II but for the other approaches, the false positives and thus the sizes of model from step I to step II are expanding. Two-step sparse boosting yields smallest bias for the coefficients estimation among the competing methods. The refined estimates after incorporating the within-subject correlation generally perform better than the initial estimates without taking into account of the within-subject correlation since the two-step methods gain reduction of bias, especially when the within-subject correlation is high. In other words, the reduction of bias from step I to step II are much larger when the within-subject correlation is higher. This is intuitive as in the second step, the within-subject correlation structure estimated from the first step have been taken into account. The similar results obtained for the bias of the estimated covariance matrix. The bias under smaller within-subject correlation is smaller than under larger within-subject correlation. The two-step sparse boosting yields smaller bias of the estimated covariance matrix than other competing methods when the within-subject correlation is high. In summary, the performance of variable selection and functional coefficients estimation for two-step sparse boosting is quite satisfactory.

3.3 Yeast cell cycle gene expression data analysis

The cell cycle is one of the most important activities in life by which cells grow, replicate their chromosomes, undergo mitosis, and split into daughter cells. Thus, identifying cell cycle-regulated genes becomes very important. Adopting a model-based approach, Luan and Li [37] identified $n = 297$ cell cycle-regulated genes based on the α-factor synchronization experiments. All gene expression levels were measured at $m = 18$ different time points covering two cell-cycle periods. Using the same subset of the original data as in [38], a total $p = 96$ transcriptional factors (TFs) are included as predictors in the downstream analysis. Wei, Huang and Li [39] proved that the effects of the TFs on gene expression levels are time-dependent. After the independence screening by l^2-norm [40] to screen out the irrelevant predictors at first step, several methods can be used to identify the key TFs involved in gene regulation. Except two-step L2 boosting and two-step sparse boosting which take into account of the within-subject correlation in the second step, one-step L2 boosting and one-step sparse boosting which ignore the within-subject correlation are also considered for better comparison. Besides, some two-step penalized approaches are also considered: two-step lasso, two-step adaptive lasso and two-step elastic net (the elastic net mixing parameter 0.5).

The results from [18] show that boosting approaches yield sparser model than the penalized methods. Sparse boosting yields even sparser model and smaller errors in terms of estimation and prediction than L2 boosting. Two-step boosting achieves better performance than one-step boosting with smaller estimation and prediction errors. Two-step sparse boosting method yields the most sparse model, with the smallest in-sample and out-of-sample prediction errors compared to other methods. In terms of the selected TFs, there is a significant overlap between two-step sparse boosting and each of the other methods. In conclusion, the two-step sparse boosting approach performs quite well in terms of variable selection, coefficients estimation and prediction and can provide useful information in identifying the important TFs that take part in the network of regulations.

4. Multi-step sparse boosting for subgroup identification

As personalized medicine is gaining popularity, identification of subgroups of the patients that can gain a higher efficacy from the treatment becomes greatly important. Recently, significant statistical approaches have been proposed to identify subgroups of patients who may be suitable for different treatments. Traditionally, subgroup identification is achieved by parametric partitioning approaches such as Bayesian approaches [41] or classification and regression tree (CART) [42]. Recently, recursive partitioning methods gain popularity since they achieve greater generalizability and efficiency. Such methods include MOB [43], PRIM [44], sequential-BATTing [45] and other non-parametric methods. For a detailed literature review of subgroup identification refer to Lipkovich et al. [46]. In this section, a sparse boosting based subgroup identification method is presented in the context of dense longitudinal data.

In particular, a formal subgroup identification method for high-dimensional dense longitudinal data is presented. It incorporates multi-step sparse boosting into the homogeneous pursuit via change point detection. Firstly, sparse boosting algorithm for individual modeling is first performed to obtain initial estimates. Then, change point detection via binary segmentation is used to identify the subgroup structure of patients. Lastly, the model on each identified subgroups is refitted and again sparse boosting is utilized to remove irrelevant predictors and yield reliable final estimates. The rest of the section is organized as follows. In Section 4.1, the subgroup model is formulated and a detailed method for subgroup identification and estimation is presented. In Section 4.2, the subgroup identification technique is evaluated through simulation studies. In Section 4.3, the feasibility and applicability of the approach is validated by studying a wallaby growth dataset.

4.1 Methodology

4.1.1 Patients model

Denote Y_{it} be the continuous measurement of the tth follow-up for patient i, where $i = 1, \cdots, n, t = 1, \cdots, T_i$. Let $\mathbf{X}_{it} = (X_{it,1}, \cdots, X_{it,p})$ be the corresponding p-dimensional predictors. Assume n patients are independent. The following longitudinal model for the patients is considered:

$$Y_{it} = \tilde{\beta}_{i,0} + \sum_{j=1}^{p} X_{it,j}\tilde{\beta}_{i,j} + \varepsilon_{it}, \quad i = 1, \cdots, n, \quad t = 1, \cdots, T_i. \tag{24}$$

where $\varepsilon_i = (\varepsilon_{i1}, \cdots, \varepsilon_{iT_i})^T, i = 1, \cdots, n$ are multivariate error terms with mean zero. Errors are assumed to be uncorrelated for different i, but components of ε_i are correlated with each other.

Moreover, the model is further assumed to have the following subgroup structure:

$$\tilde{\beta}_{i,j} = \begin{cases} \beta_{1,j} & when \ \ i \in \Omega_{1,j} \\ \beta_{2,j} & when \ \ i \in \Omega_{2,j} \\ \vdots & \vdots \\ \beta_{\mathcal{N}_j+1,j} & when \ \ i \in \Omega_{\mathcal{N}_j+1,j} \end{cases} \tag{25}$$

The partition for regression coefficient $\left\{\tilde{\beta}_{i,j} : \ 1 \le i \le n\right\}$ is $\{\Omega_{k,j} : \ 1 \le k \le \mathcal{N}_j + 1\}$, which is unknown, and thus there are $\mathcal{N}_j + 1$ subgroups for the jth predictor. All patients are divided into at least $\max_j(\mathcal{N}_j + 1)$ and at most $\prod_{j=0}^{p}(\mathcal{N}_j + 1)$ subgroups by the model. The patients in the same subgroup share a similar relationship between the response and the predictors and have the same set of regression coefficients while different subgroups have different overall relationship between response and covariates. The main aim is to investigate the effects of the predictors on the response for different subgroups.

However, if the number of predictors under consideration is much larger than the number of patients and the number of follow-ups, a serious challenge may arise to estimate regression coefficients. Therefore, instead of adopting traditional methods (eg, MLE), sparse boosting method can be used to estimate the regression coefficients. With this, the dimensionality of features can be reduced and the coefficients of parameters can be obtained simultaneously.

4.1.2 Subgroup identification and estimation

Denote $\tilde{\beta}_i = \left(\tilde{\beta}_{i,0}, \ \cdots, \ \tilde{\beta}_{i,p}\right)^T$ and $\tilde{\beta} = \left(\tilde{\beta}_1^T, \ \cdots, \ \tilde{\beta}_n^T\right)^T$. Firstly, an initial estimator for $\tilde{\beta}_i$ is calculated for each subject i through sparse boosting approach using his or her own repeated measurements data; then, homogeneity pursuit via change point detection can be used to identify the change points among $\beta_{k,j}$s; lastly, the $\tilde{\beta}_i$s can be replaced by the identified subgroup structure, and the final estimator of regression coefficients can be obtained by the sparse boosting algorithm again. The steps for estimating $\tilde{\beta}_i$ is outlined as below.

In the first step, individualized modeling via sparse boosting is performed. For each of the ith individual, the initial coefficients $\tilde{\beta}_i$ can be estimated by minimizing the following least squares loss function:

$$\sum_{t=1}^{T_i}\left(Y_{it} - \tilde{\beta}_{i,0} - \sum_{j=1}^{p} X_{it,j}\tilde{\beta}_{i,j}\right)^2. \tag{26}$$

Let $Y_i = (Y_{i1}, \ \cdots, \ Y_{iT_i})^T, X_{it,0} = 1, X_{i,j} = \left(X_{i1,j}, \ \cdots, \ X_{iT_i,j}\right)^T, \mathbf{X}_i = (X_{i,0}, \ \cdots, \ X_{i,p})$. Then the function Eq. (26) can be written in the matrix form:

$$\left(Y_i - \mathbf{X}_i\tilde{\beta}_i\right)^T\left(Y_i - \mathbf{X}_i\tilde{\beta}_i\right). \tag{27}$$

Denote $\tilde{\beta}_i^{[\hat{L}_i]} = \left(\tilde{\beta}_{i,0}^{[\hat{L}_i]} \ \cdots, \ \tilde{\beta}_{i,p}^{[\hat{L}_i]}\right)^T$ to be the estimator of $\tilde{\beta}_i$ by sparse boosting with Eq. (27) being loss function, where $\hat{L}_i$ is the estimated stopping iterations in this step. This is the initial estimator of $\tilde{\beta}_i$. The detailed sparse boosting algorithm will be presented in the next subsection.

In the second step, homogeneity pursuit via change point detection is performed. Binary segmentation algorithm [47] is used to detect the change points among $\tilde{\beta}_{i,j}$, $i = 1, \cdots, n$ and to identify the subgroup structure. Let $\tilde{\beta}_{i,j}^{[\hat{L}_i]}$ be the $(j+1)$th component of $\tilde{\beta}_i^{[\hat{L}_i]}$. For the jth covariate, $\tilde{\beta}_{i,j}^{[\hat{L}_i]}$, $i = 1, \cdots, n$, are sorted in ascending order, and denoted by $b_{(1)} \leq \cdots \leq b_{(n)}$. Denote $r_{i,j}$ be the rank of $\tilde{\beta}_{i,j}^{[\hat{L}_i]}$.

For any $1 \leq l_1 < l_2 \leq n$, denote the scaled difference between the partial means of the first $\tau - l_1 + 1$ observations and the last $l_2 - \tau$ observations to be

$$H_{l_1 l_2}(\tau) = \sqrt{\frac{(l_2 - \tau)(\tau - l_1 + 1)}{l_2 - l_1 + 1}} \left(\frac{\sum_{i=\tau+1}^{l_2} b_{(l)}}{l_2 - \tau} - \frac{\sum_{i=l_1}^{\tau} b_{(i)}}{\tau - l_1 + 1} \right). \tag{28}$$

Denote δ to be the threshold, which is a tuning parameter and can be selected by AIC or BIC, then the binary segmentation algorithm is as follows:

1. Find $\hat{t}_1$ such that

$$H_{1,n}(\hat{t}_1) = \max_{1 \leq \tau < n} H_{1,n}(\tau). \tag{29}$$

If $H_{1,n}(\hat{t}_1) \leq \delta$, there is no change points among $b_{(l)}$, $l = 1, \cdots, n$, and the change point detection process terminates. Otherwise, $\hat{t}_1$ is added to the set of change points and the region $\{\tau : 1 \leq \tau \leq n\}$ is divided into two subregions: $\{\tau : 1 \leq \tau \leq \hat{t}_1\}$ and $\{\tau : \hat{t}_1 + 1 \leq \tau \leq n\}$.

2. Find the change points in the two subregions derived in part (1), respectively. Consider the region $\{\tau : 1 \leq \tau \leq \hat{t}_1\}$ first. Find $\hat{t}_2$ such that

$$H_{1,\hat{t}_1}(\hat{t}_2) = \max_{1 \leq \tau < \hat{t}_1} H_{1,\hat{t}_1}(\tau). \tag{30}$$

If $H_{1,\hat{t}_1}(\hat{t}_2) \leq \delta$, there is no change point in the region $\{\tau : 1 \leq \tau \leq \hat{t}_1\}$. Otherwise, add $\hat{t}_2$ to the set of change points and divide the region $\{\tau : 1 \leq \tau \leq \hat{t}_1\}$ into two subregions: $\{\tau : 1 \leq \tau \leq \hat{t}_2\}$ and $\{\tau : \hat{t}_2 + 1 \leq \tau \leq \hat{t}_1\}$. Similarly, for the region $\{\tau : \hat{t}_1 + 1 \leq \tau \leq n\}$, $\hat{t}_3$ can be found such that

$$H_{\hat{t}_1+1,n}(\hat{t}_3) = \max_{\hat{t}_1+1 \leq \tau < n} H_{\hat{t}_1+1,n}(\tau). \tag{31}$$

If $H_{\hat{t}_1+1,n}(\hat{t}_3) \leq \delta$, there is no change point in the region $\{\tau : \hat{t}_1 + 1 \leq \tau \leq n\}$. Otherwise, add $\hat{t}_3$ to the set of change points and divide the region $\{\tau : \hat{t}_1 + 1 \leq \tau \leq n\}$ into two subregions: $\{\tau : \hat{t}_1 + 1 \leq \tau \leq \hat{t}_3\}$ and $\{\tau : \hat{t}_3 + 1 \leq \tau \leq n\}$.

3. For each subregion derived in part (2), the above algorithm is repeated for the subregion $\{\tau : 1 \leq \tau \leq \hat{t}_1\}$ or $\{\tau : \hat{t}_1 + 1 \leq \tau \leq n\}$ in part (2) until no change point is detected in any subregions.

The estimated locations for change points are sorted in increasing order and denoted by

$$\hat{t}_{(1)} < \hat{t}_{(2)} < \cdots < \hat{t}_{(\hat{\mathcal{N}}_j)}, \tag{32}$$

where $\hat{\mathcal{N}}_j$ is the number of detected change points and could be used to estimate $\mathcal{N}_j$. Further denote $\hat{t}_{(0)} = 0$, and $\hat{t}_{(\hat{\mathcal{N}}_j+1)} = n$. Let. $\hat{R}_{i,j} = \{\ell : \hat{t}_{(\ell-1)} < r_{i,j} \leq \hat{t}_{(\ell)}\}$, $1 \leq \ell \leq \hat{\mathcal{N}}_{j+1}$, where $\{\hat{R}_{i,j} : 1 \leq i \leq n\}$ can be used to estimate the grouping index $\{R_{i,j} : 1 \leq i \leq n\}$. The above algorithm can be used to identify the change points for all $j = 0, \cdots, p$ and correspondingly obtain $\{\hat{R}_{i,j} : 1 \leq i \leq n, 0 \leq j \leq p\}$. Let $\left\{\hat{R}^{\star}_{\ell,j} : 1 \leq \ell \leq \hat{\mathcal{N}}, 0 \leq j \leq p\right\}$ = unique rows of $\{\hat{R}_{i,j} : 1 \leq i \leq n, 0 \leq j \leq p\}$, then $\hat{\mathcal{N}}$ is the estimated total number of subgroups for patients and the patients index in group ℓ is.

$$\hat{\Omega}_{\ell} = \left\{i : \hat{R}_{i,j} = \hat{R}^{\star}_{\ell,j}\right\}, \quad 1 \leq \ell \leq \hat{\mathcal{N}}. \tag{33}$$

All the coefficients $\tilde{\beta}_{i,j}$s in the same estimated subgroup $\hat{\Omega}_{\ell}$ are treated to be equal.

In the third step, subgroup modeling is performed by sparse boosting. Incorporating the patients structure identified in step 2, the model is refitted to each of the subgroups via sparse boosting with the following least squares loss function

$$\sum_{i \in \hat{\Omega}_{\ell}} \sum_{t=1}^{T_i} \left(Y_{it} - \tilde{\beta}_{i,0} - \sum_{j=1}^{p} X_{it,j}\tilde{\beta}_{i,j}\right)^2, \quad 1 \leq \ell \leq \hat{\mathcal{N}}. \tag{34}$$

Further denote $Y^{\star}_{\ell} = \left(Y^T_{\hat{\Omega}_{\ell}[1]}, \cdots, Y^T_{\hat{\Omega}_{\ell}[|\hat{\Omega}_{\ell}|]}\right)^T, X^{\star}_{\ell,j} = \left(X^T_{\hat{\Omega}_{\ell}[1],j}, \cdots, X^T_{\hat{\Omega}_{\ell}[|\hat{\Omega}_{\ell}|],j}\right)^T$, $\mathbf{X}^{\star}_{\ell} = \left(X^{\star}_{\ell,0}, \cdots, X^{\star}_{\ell,p}\right)$ and $\tilde{\beta}^{\star}_{\ell} = \left(\tilde{\beta}^T_{\hat{\Omega}_{\ell}[1]}, \cdots, \tilde{\beta}^T_{\hat{\Omega}_{\ell}[|\hat{\Omega}_{\ell}|]}\right)^T$ for $\ell = 1, \cdots, \hat{\mathcal{N}}$, where $\hat{\Omega}_{\ell}[i]$ is the ith element of $\hat{\Omega}_{\ell}$ and $|\hat{\Omega}_{\ell}|$ is the number of elements in $\hat{\Omega}_{\ell}$. The function Eq. (34) can be written in the matrix form:

$$\left(Y^{\star}_{\ell} - \mathbf{X}^{\star}_{\ell}\tilde{\beta}^{\star}_{\ell}\right)^T \left(Y^{\star}_{\ell} - \mathbf{X}^{\star}_{\ell}\tilde{\beta}^{\star}_{\ell}\right), \quad 1 \leq \ell \leq \hat{\mathcal{N}}. \tag{35}$$

Denote $\tilde{\beta}^{\star\left[\hat{L}^{\star}_{\ell}\right]}_{\ell}$ to be the estimate for $\tilde{\beta}^{\star}_{\ell}$ by sparse boosting with Eq. (35) being the loss function, where $\hat{L}^{\star}_{\ell}$ is the estimated number of stopping iterations in this step. The estimator for coefficient $\tilde{\beta}_i$ is

$$\hat{\beta}_i = \left\{\tilde{\beta}^{\star\left[\hat{L}^{\star}_{\ell}\right]}_{\ell} \quad \text{for}\, i \in \hat{\Omega}_{\ell}\right\}, \quad 1 \leq i \leq n. \tag{36}$$

More details about how to use sparse boosting to obtain $\left\{\tilde{\beta}^{[\hat{L}_i]}_i, 1 \leq i \leq n\right\}$ and $\left\{\tilde{\beta}^{\star\left[\hat{L}^{\star}_{\ell}\right]}_{\ell}, 1 \leq \ell \leq \hat{\mathcal{N}}\right\}$ are given in the following subsection.

4.1.3 Multi-step sparse boosting techniques

gMDL can be used as the penalized empirical risk function to estimate the update criterion in each iteration and the stopping criterion to avoid the selection of the tuning parameter. gMDL can be expressed in the following form:

$$\text{gMDL}(Y, RSS, \text{trace}(\mathcal{B})) = \log(F) + \frac{\text{trace}(\mathcal{B})}{|Y|}\log\left(\frac{Y^TY - RSS}{\text{trace}(\mathcal{B}) \times F}\right),$$
$$F = \frac{RSS}{|Y| - \text{trace}(\mathcal{B})}, \tag{37}$$

where Y is the vector of response variable, $|Y|$ is the length of Y, $\mathcal{B}$ is the boosting operator and RSS is the residual sum of squares.

The sparse boosting procedure is described in details. The starting value of $\tilde{\beta}_i$ is set to zero vector, i.e. $\tilde{\beta}_i^{[0]} = 0$, and in each of the l_ith iteration $(0 < l_i \le L_i$, and L_i is the maximum number of iterations considered in this step), the residual $R^{[l_i]} = Y_i - \mathbf{X}_i\tilde{\beta}_i^{[l_i-1]}$ in present iteration is used to fit each of the jth element $X_{i,j}, j = 0, \cdots, p$. The fit denoted by $\hat{\lambda}_j^{[l_i]}$ can be obtained by minimizing the squared loss function $\left(R^{[l_i]} - X_{i,j}\lambda\right)^T\left(R^{[l_i]} - X_{i,j}\lambda\right)$ with respect to λ. Thus, the least squares estimate is $\hat{\lambda}_j^{[l_i]} = \left[\left(X_{i,j}\right)^T\left(X_{i,j}\right)\right]^{-1}\left(X_{i,j}\right)^T R^{[l_i]}$, the corresponding hat matrix is $\mathcal{H}_j = \left(X_{i,j}\right)\left[\left(X_{i,j}\right)^T\left(X_{i,j}\right)\right]^{-1}\left(X_{i,j}\right)^T$ and the residual sum of squares is $RSS_j^{[l_i]} = \left(R^{[l_i]} - X_{i,j}\hat{\lambda}_j^{[l_i]}\right)^T\left(R^{[l_i]} - X_{i,j}\hat{\lambda}_j^{[l_i]}\right)$. The selected entry $\hat{s}_{l_i}$ is obtained by:

$$\hat{s}_{l_i} = \text{argmin}_{0 \le j \le p}\text{gMDL}\left(Y_i, RSS_j^{[l_i]}, \text{trace}\left(\mathcal{B}_j^{[l_i]}\right)\right), \tag{38}$$

where $\mathcal{B}_j^{[1]} = \mathcal{H}_j$ and $\mathcal{B}_j^{[l_i]} = I - \left(I - \mathcal{H}_j\right)\left(I - \nu\mathcal{H}_{\hat{s}_{l_i-1}}\right)\cdots\cdot(I - \nu\mathcal{H}_{\hat{s}_1})$ for $l_i > 1$ is the boosting operator for choosing jth entry in the l_ith iteration in this step. Hence, there is an unique element $X_{i,\hat{s}_{l_i}}$ to be selected at each iteration, and only the corresponding coefficient vector $\tilde{\beta}_{i,\hat{s}_{l_i}}^{[l_i]}$ changes, i.e., $\tilde{\beta}_{i,\hat{s}_{l_i}}^{[l_i]} = \tilde{\beta}_{i,\hat{s}_{l_i}}^{[l_i-1]} + \nu\hat{\lambda}_{\hat{s}_{l_i}}^{[l_i]}$, where ν is the pre-specified step-size parameter. All the other $\tilde{\beta}_{i,j}^{[l_i]}$ for $j \ne \hat{s}_{l_i}$ keep unchanged. This procedure is repeated for L_i times and the number of iterations L_i can be estimated by

$$\hat{L}_i = \text{argmin}_{1 \le l_i \le L_i}\text{gMDL}\left(Y_i, RSS_{\hat{s}_{l_i}}^{[l_i]}, \text{trace}\left(\mathcal{B}^{[l_i]}\right)\right), \tag{39}$$

where $\mathcal{B}^{[l_i]} = I - \left(I - \nu\mathcal{H}_{\hat{s}_{l_i}}\right)\cdots\cdot(I - \nu\mathcal{H}_{\hat{s}_1})$.

From the above sparse boosting approach, the estimator of $\tilde{\beta}_i$ is $\tilde{\beta}_i^{[\hat{L}_i]} = \left(\tilde{\beta}_{i,0}^{[\hat{L}_i]}, \cdots, \tilde{\beta}_{i,p}^{[\hat{L}_i]}\right)^T, i = 1, \cdots, n$. Then the subgroup structure can be obtained by homogeneity pursuit via change point detection.

Next, sparse boosting is used again for each estimated subgroups. The starting value of $\tilde{\beta}_\ell^\star$ is set to zero vector, i.e. $\tilde{\beta}_\ell^{\star[0]} = 0$, and in each of the $l_\ell^\star$th iteration $(0 < l_\ell^\star \le L_\ell^\star$, and $L_\ell^\star$ is the maximum number of iterations considered in this stage), the residual $R^{\star[l_\ell^\star]} = Y_\ell^\star - \mathbf{X}_\ell^\star\tilde{\beta}_\ell^{\star[l_\ell^\star-1]}$ in present iteration is used to fit each of the jth component $X_{\ell,j}^\star, j = 0, \cdots, p$. Then the fit denoted by $\hat{\lambda}_j^{\star[l_i^\star]}$ can be calculated by minimizing the squared loss function $\left(R^{\star[l_\ell^\star]} - X_{\ell,j}^\star\lambda\right)^T\left(R^{\star[l_\ell^\star]} - X_{i,j}^\star\lambda\right)$ with respect to λ. Therefore, the least squares estimate is $\hat{\lambda}_j^{\star[l_\ell^\star]} = \left[\left(X_{\ell,j}^\star\right)^T\left(X_{\ell,j}^\star\right)\right]^{-1}\left(X_{\ell,j}^\star\right)^T R^{\star[l_\ell^\star]}$, the corresponding hat matrix is $\mathcal{H}_j^\star = \left(X_{\ell,j}^\star\right)\left[\left(X_{\ell,j}^\star\right)^T\left(X_{\ell,j}^\star\right)\right]^{-1}\left(X_{\ell,j}^\star\right)^T$ and the residual sum of squares is $RSS_j^{\star[l_\ell^\star]} = \left(R^{\star[l_\ell^\star]} - X_{\ell,j}^\star\hat{\lambda}_j^{[l_\ell^\star]}\right)^T\left(R^{\star[l_\ell^\star]} - X_{\ell,j}^\star\hat{\lambda}_j^{[l_\ell^\star]}\right)$. The chosen element $\hat{s}_{l_\ell^\star}^\star$ is attained by:

$$\hat{s}_{l_\ell^\star}^\star = \text{argmin}_{0\le j\le p}\text{gMDL}\left(Y_\ell^\star, RSS_j^{\star[l_\ell^\star]}, \text{trace}\left(\mathcal{B}_j^{\star[l_\ell^\star]}\right)\right), \tag{40}$$

where $\mathcal{B}_j^{\star[1]} = \mathcal{H}_j^\star$ and $\mathcal{B}_j^{\star[l_\ell]} = I - \left(I - \mathcal{H}_j^\star\right)\left(I - \nu\mathcal{H}_{\hat{s}_{l_\ell^\star-1}}^\star\right)\cdots\left(I - \nu\mathcal{H}_{\hat{s}_1}^\star\right)$ for $l_\ell^\star > 1$ is the boosting operator for choosing jth element in the $l_\ell^\star$th iteration in this stage. Hence, there is an unique element $X_{\ell,\hat{s}_{l_\ell^\star}}^\star$ to be selected at each iteration, and only the corresponding coefficient vector $\tilde{\beta}_{\ell,\hat{s}_{l_\ell^\star}}^{\star[l_\ell^\star]}$ changes, i.e., $\tilde{\beta}_{\ell,\hat{s}_{l_\ell^\star}}^{[l_\ell]} = \tilde{\beta}_{\ell,\hat{s}_{l_\ell^\star}}^{\star[l_\ell^\star-1]} + \nu\hat{\lambda}_{\hat{s}_{l_\ell^\star}}^{[l_\ell^\star]}$, where ν is the pre-specified step-size parameter. All the other $\tilde{\beta}_{\ell,j}^{\star[l_\ell^\star]}$ for $j \neq \hat{s}_{l_\ell^\star}$ keep unchanged. This procedure is repeated for $L_\ell^\star$ times and the number of iterations $L_\ell^\star$ can be estimated by

$$\hat{L}_i^\star = \text{argmin}_{1\le l_\ell\le L_i^\star}\text{gMDL}\left(Y_\ell^\star, RSS_{\hat{s}_{l_\ell^\star}}^{\star[l_\ell^\star]}, \text{trace}\left(\mathcal{B}^{\star[l_\ell^\star]}\right)\right), \tag{41}$$

where $\mathcal{B}^{\star[l_\ell^\star]} = I - \left(I - \nu\mathcal{H}_{\hat{s}_{l_\ell^\star}}^\star\right)\cdots\left(I - \nu\mathcal{H}_{\hat{s}_1}^\star\right)$.

From the second step of sparse boosting, the estimator of $\tilde{\beta}_\ell$ is $\tilde{\beta}_\ell^{\star[\hat{L}_\ell^\star]} = \left(\tilde{\beta}_{\ell,0}^{\star[\hat{L}_\ell^\star]}, \cdots, \tilde{\beta}_{\ell,p}^{\star[\hat{L}_\ell^\star]}\right)^T, \ell = 1, \cdots, \hat{\mathcal{N}}$.

4.2 Simulation

Extensive simulations are conducted to evaluate the performance of the proposed procedure. The accuracy of subgrouping, feature selection, coefficients estimation and prediction are assessed in the setting of different number of patients and repeated measurements. To understand the advantage of the proposed method better, the following four approaches are also considered. M1: the homogeneous model fitting method which treats all patients as one group and use sparse boosting for the single model to estimate $\tilde{\beta}$; M2: the heterogeneous model fitting method which uses initial pre-grouping estimate $\tilde{\beta}_i^{[\hat{L}_i]}$ as the final estimate of $\tilde{\beta}_i$; M3: same as the proposed method but in step 2, instead of detecting the change points for coefficients of each covariate $\tilde{\beta}_{i,j}^{[\hat{L}_i]}, i = 1, \cdots, n$ for $j = 0, \cdots, p$, it detects the change points among $\left(\tilde{\beta}_1^{T[\hat{L}_1]}, \cdots, \tilde{\beta}_n^{T[\hat{L}_n]}\right)^T$ similarly to Ke et al. [48]; M4: the proposed method.

The results from [19] show that the naive homogeneous model fitting method M1 can rarely identify the important covariates while the over-parameterized model fitting method M2 and other two methods (M3 & M4) which identify subgroup structures consistently yield true positives equal to the true number of important covariates. Compared these three methods which can identify the important covariates, the proposed method produces smallest false positives. In addition, the number of false positives is decreasing when there is an increase in cluster size. Neither the homogeneous model fitting method nor heterogeneous model fitting method is able to identify the true structure among patients. The method M3 produces much more subgroups than it really has, while the proposed method M4 identified the number of subgroups closest to the actual number of subgroups. Furthermore, the probability of identifying the true subgroups becomes larger when the number of repeated measurements increases. For in-sample prediction, the over-parameterized model M2 performs the best while the methods M3 & M4 performs very competitively. However, for out-of-sample prediction, method M4 is the best. M1 is inferior to M4, yielding poor results of estimation and prediction. In summary, the proposed method preforms pretty well in terms of subgroup identification, variable selection, estimation as well as prediction.

4.3 Wallaby growth data analysis

The proposed subgroup identification method is applied to wallaby growth data, which is from the Australian Dataset and Story Library (OzDASL) and can be found at http://www.statsci.org/data/oz/wallaby.html. The data set has 77 Tammar wallabies' growth measurements which were taken longitudinally. The response variable is the weight of wallabies (tenths of a gram). The predictors involve length of head, ear, body, arm, leg, tail, foot and their second order interactions. Therefore, a total of 35 predictors are included in the analysis. After removing the missing data, 43 Tammar wallabies are kept in our dataset. The number of repeated measurements ranges from 9 to 34 (median: 23). To have a better understanding of the wallabies' growth trend, the questions of which parts of body would affect the weight and whether the length of each body parts have the same effects on the weight for all wallabies are investigated, i.e. is there any subgroups among wallabies. Except the above subgroup identification method (SB-CPD1), the other 3 methods studied in simulation are also considered, i.e. homogeneous model fitting method (SB-Homogeneous), heterogeneous model fitting method (SB-Heterogeneous) and the method similar to SB-CPD1 but identifying subgroups via other method in Ke et al. [48] (SB-CPD2). In addition, the following subgroup identification methods incorporating penalized methods are also investigated: similar to our proposed method but instead of using sparse boosting, lasso (Lasso-CPD1), elastic net (ElasticNet-CPD1), SCAD (SCAD-CPD1) or MCP (MCP-CPD1) is used.

The results from [19] show that although Lasso-CPD1 and ElasticNet-CPD1 yield smaller in-sample prediction error by keeping all 35 covariates, they have relatively large out-of-sample prediction errors due to over-fitting problem. The subgroup identification method via sparse boosting keeps smaller number of predictors, achieves sparser model than penalized methods. The proposed method SB-CPD1 identifies smaller number of subgroups and predictors than alternative competing methods while produces smallest out-of-sample prediction errors. In conclusion, the proposed subgroup identification method provides a more precise definition for various subgroups. It may also result in a more accurate medical decision making for these subjects.

5. Conclusions

In this chapter, we discussed various sparse boosting based machine learning methods in the context of high-dimensional data problems. Specifically, we presented the sparse boosting procedure and two-step sparse boosting procedure for nonparametric varying-coefficient models with survival data and repeatedly measured longitudinal data respectively to simultaneously perform variable selection and estimation of functional coefficients. We further presented the multi-step sparse boosting based subgroup identification method with longitudinal patient data to identify subgroups that exhibit different treatment effects. The extensive numerical studies show the validity and effectiveness of our proposed methods and the real data analysis further demonstrate their usefulness and advantages.

Author details

Mu Yue
Singapore University of Technology and Design, Singapore

*Address all correspondence to: yuemu.moon@gmail.com

References

[1] Tibshirani R. Regression shrinkage and selection via the lasso. Journal of the Royal Statistical Society: Series B (Methodological). 1996 Jan;58(1):267–288.

[2] Fan J, Li R. Variable selection via nonconcave penalized likelihood and its oracle properties. Journal of the American statistical Association. 2001 Dec 1;96(456):1348–1360.

[3] Zhang CH. Nearly unbiased variable selection under minimax concave penalty. The Annals of statistics. 2010; 38(2):894–942.

[4] Zou H. The adaptive lasso and its oracle properties. Journal of the American statistical association. 2006 Dec 1;101(476):1418–1429.

[5] Zou H, Hastie T. Regularization and variable selection via the elastic net. Journal of the royal statistical society: series B (statistical methodology). 2005 Apr 1;67(2):301–320.

[6] Yuan M, Lin Y. Model selection and estimation in regression with grouped variables. Journal of the Royal Statistical Society: Series B (Statistical Methodology). 2006 Feb;68(1):49–67.

[7] Schapire RE. The strength of weak learnability. Machine learning. 1990 Jun 1;5(2):197–227.

[8] Freund Y, Schapire RE. A decision-theoretic generalization of on-line learning and an application to boosting. Journal of computer and system sciences. 1997 Aug 1;55(1):119–139.

[9] Bühlmann P, Yu B. Boosting with the L 2 loss: regression and classification. Journal of the American Statistical Association. 2003 Jun 1;98(462):324–339.

[10] Bühlmann P, Yu B, Singer Y, Wasserman L. Sparse Boosting. Journal of Machine Learning Research. 2006 Jun 1;7(6).

[11] Wang Z. HingeBoost: ROC-based boost for classification and variable selection. The International Journal of Biostatistics. 2011 Feb 4;7(1).

[12] Bühlmann P, Hothorn T. Twin boosting: improved feature selection and prediction. Statistics and Computing. 2010 Apr;20(2):119–138.

[13] Komori O, Eguchi S. A boosting method for maximizing the partial area under the ROC curve. BMC bioinformatics. 2010 Dec;11(1):1–7.

[14] Wang Z. Multi-class hingeboost. Methods of information in medicine. 2012;51(02):162–167.

[15] Zhao J. General sparse boosting: improving feature selection of l2 boosting by correlation-based penalty family. Communications in Statistics-Simulation and Computation. 2015 Jul 3; 44(6):1612–1640.

[16] Yang Y, Zou H. Nonparametric multiple expectile regression via ER-Boost. Journal of Statistical Computation and Simulation. 2015 May 3;85(7):1442–1458.

[17] Yue M, Li J, Ma S. Sparse boosting for high-dimensional survival data with varying coefficients. Statistics in medicine. 2018 Feb 28;37(5):789–800.

[18] Yue M, Li J, Cheng MY. Two-step sparse boosting for high-dimensional longitudinal data with varying coefficients. Computational Statistics Data Analysis. 2019 Mar 1;131:222–234.

[19] Yue M, Huang L. A new approach of subgroup identification for high-dimensional longitudinal data. Journal of Statistical Computation and

Simulation. 2020 Jul 23;90(11):2098–2116.

[20] David CR. Regression models and life tables (with discussion). Journal of the Royal Statistical Society. 1972;34(2): 187–220.

[21] Wei LJ. The accelerated failure time model: a useful alternative to the Cox regression model in survival analysis. Statistics in medicine. 1992;11(14–15): 1871–1879.

[22] Schmid M, Hothorn T. Flexible boosting of accelerated failure time models. BMC bioinformatics. 2008 Dec; 9(1):1–3.

[23] Wang Z, Wang CY. Buckley-James boosting for survival analysis with high-dimensional biomarker data. Statistical Applications in Genetics and Molecular Biology. 2010 Jun 8;9(1).

[24] Li J, Ma S. Survival analysis in medicine and genetics. CRC Press; 2013 Jun 4.

[25] Stute W. Consistent estimation under random censorship when covariables are present. Journal of Multivariate Analysis. 1993 Apr 1;45(1): 89–103.

[26] Curry HB, Schoenberg IJ. On Pólya frequency functions IV: the fundamental spline functions and their limits. InIJ Schoenberg Selected Papers 1988 (pp. 347–383). Birkhäuser, Boston, MA.

[27] Hansen MH, Yu B. Model selection and the principle of minimum description length. Journal of the American Statistical Association. 2001 Jun 1;96(454):746–774.

[28] Shedden K, Taylor JM, Enkemann SA, Tsao MS, Yeatman TJ, Gerald WL, Eschrich S, Jurisica I, Giordano TJ, Misek DE, Chang AC. Gene expression–based survival prediction in lung adenocarcinoma: a multi-site, blinded validation study. Nature medicine. 2008 Aug;14(8):822.

[29] Consonni D, Bertazzi PA, Zocchetti C. Why and how to control for age in occupational epidemiology. Occupational and environmental medicine. 1997 Nov 1;54(11):772–776.

[30] Wang S, Nan B, Zhu J, Beer DG. Doubly penalized Buckley–James method for survival data with high-dimensional covariates. Biometrics. 2008 Mar;64(1):132–140.

[31] Diggle P, Heagerty P, Liang KY, Zeger S. Analysis of Longitudinal Data: Oxford University Press. 2002.

[32] Fitzmaurice GM, Laird NM, Ware JH. Applied longitudinal analysis, vol. 998 John Wiley & Sons. Hoboken NJ. 2012.

[33] Lin X, Carroll RJ. Semiparametric regression for clustered data using generalized estimating equations. Journal of the American Statistical Association. 2001 Sep 1;96(455):1045–1056.

[34] Fan J, Huang T, Li R. Analysis of longitudinal data with semiparametric estimation of covariance function. Journal of the American Statistical Association. 2007 Jun 1;102(478):632–641.

[35] Cheng MY, Honda T, Li J, Peng H. Nonparametric independence screening and structure identification for ultra-high dimensional longitudinal data. Annals of Statistics. 2014;42(5):1819–1849.

[36] Cheng MY, Honda T, Li J. Efficient estimation in semivarying coefficient models for longitudinal/clustered data. The Annals of Statistics. 2016;44(5): 1988–2017.

[37] Luan Y, Li H. Clustering of time-course gene expression data using a mixed-effects model with B-splines.

Bioinformatics. 2003 Mar 1;19(4): 474–482.

[38] Wang L, Chen G, Li H. Group SCAD regression analysis for microarray time course gene expression data. Bioinformatics. 2007 Jun 15;23(12): 1486–1494.

[39] Wei F, Huang J, Li H. Variable selection and estimation in high-dimensional varying-coefficient models. Statistica Sinica. 2011 Oct 1;21(4):1515.

[40] Yue M, Li J. Improvement screening for ultra-high dimensional data with censored survival outcomes and varying coefficients. The international journal of biostatistics. 2017 May 18;13(1).

[41] Sivaganesan S, MÃ¼ller P, Huang B. Subgroup finding via Bayesian additive regression trees. Statistics in medicine. 2017 Jul 10;36(15):2391–2403.

[42] Zhang H, Singer BH. Recursive partitioning and applications. Springer Science & Business Media; 2010 Jul 1.

[43] Zeileis A, Hothorn T, Hornik K. Model-based recursive partitioning. Journal of Computational and Graphical Statistics. 2008 Jun 1;17(2):492–514.

[44] Chen G, Zhong H, Belousov A, Devanarayan V. A PRIM approach to predictive-signature development for patient stratification. Statistics in medicine. 2015 Jan 30;34(2):317–342.

[45] Huang X, Sun Y, Trow P, Chatterjee S, Chakravartty A, Tian L, Devanarayan V. Patient subgroup identification for clinical drug development. Statistics in medicine. 2017 Apr 30;36(9):1414–1428.

[46] Lipkovich I, Dmitrienko A, B D'Agostino Sr R. Tutorial in biostatistics: data-driven subgroup identification and analysis in clinical trials. Statistics in medicine. 2017 Jan 15; 36(1):136–196.

[47] Bai J. Estimating multiple breaks one at a time. Econometric theory. 1997 Jun 1:315–352.

[48] Ke Y, Li J, Zhang W. Structure identification in panel data analysis. Annals of Statistics. 2016;44(3):1193–1233.

Chapter 4

Fast Computation of the EM Algorithm for Mixture Models

Masahiro Kuroda

Abstract

Mixture models become increasingly popular due to their modeling flexibility and are applied to the clustering and classification of heterogeneous data. The EM algorithm is largely used for the maximum likelihood estimation of mixture models because the algorithm is stable in convergence and simple in implementation. Despite such advantages, it is pointed out that the EM algorithm is local and has slow convergence as the main drawback. To avoid the local convergence of the EM algorithm, multiple runs from several different initial values are usually used. Then the algorithm may take a large number of iterations and long computation time to find the maximum likelihood estimates. The speedup of computation of the EM algorithm is available for these problems. We give the algorithms to accelerate the convergence of the EM algorithm and apply them to mixture model estimation. Numerical experiments examine the performance of the acceleration algorithms in terms of the number of iterations and computation time.

Keywords: the EM algorithm, normal mixture models, acceleration of convergence, the vector ε algorithm, restarting procedure, initial value selection, the emEM algorithm

1. Introduction

Mixture models become increasingly popular due to their modeling flexibility and are applied to the clustering and classification of heterogeneous data, see [1–3]. The EM algorithm [4] is largely used for the maximum likelihood estimation of mixture models because the algorithm is stable in convergence and simple in implementation. Despite such advantages, it is pointed out that the EM algorithm is local and has slow convergence as the main drawback.

To circumvent the problem of slow convergence of the EM algorithm, various acceleration algorithms incorporating optimization methods are proposed. The optimization methods include the multivariate Aitken method [5], the conjugate gradient method [6], and the quasi-Newton method [7, 8]. However, these methods require matrix computation such as matrix inversion or evaluation of Hessian and Jacobian matrices and a line search for step length optimization. Therefore, their acceleration algorithms tend to lack one or more of the nice properties of the EM algorithm, although they may converge faster than the EM algorithm.

As another approach, the ε-accelerated EM algorithm [9] is proposed to accelerate the convergence of the EM algorithm by using the vector ε (vε) algorithm [10] that is a vector extrapolation algorithm [11, 12]. The vε algorithm can accelerate the convergence of the sequence of estimates from the EM algorithm, and therefore, the ε-accelerated EM algorithm does not require any modification of the E- and M-steps of the EM algorithm. This point is the advantage of the ε-accelerated EM algorithm over other acceleration algorithms using the optimization methods. To reduce the number of iterations and computation time of the ε-accelerated EM algorithm, the εR-accelerated EM algorithm [13] is developed. The algorithm improves the computation speed of the ε-accelerated EM algorithm by embedding a restarting procedure. Then the restarting procedure finds a value for restarting the EM iterations such that a newly generated sequence of EM iterations from the value moves quickly into a neighborhood of a stationary point. We use the ε-accelerated EM and εR-accelerated EM algorithms for parameter estimation.

In application of the EM algorithm to mixture models, the algorithm is sensitive to the choice of the initial value and may find estimates at a local maximum of the log-likelihood function. Several strategies are proposed to efficiently initiate the EM algorithm for getting the global maximum of the log-likelihood function, see [14–17]. We use the emEM algorithm [14] for the mixture model estimation and improve its computation speed by the ε-accelerated EM and εR-accelerated EM algorithms.

The chapter is organized as follows. Section 2 describes the EM algorithm for normal mixture models. In Section 3, we introduce the ε-accelerated EM and εR-accelerated EM algorithms. Section 4 presents numerical experiments to evaluate the performance of these acceleration algorithms. In Section 5, we provide an acceleration algorithm that applies the ε-accelerated EM and εR-accelerated EM algorithms to the emEM algorithm. Numerical experiments in Section 6 study the effects of these acceleration algorithms on the emEM algorithm. In Section 7, we present our concluding remarks.

2. The EM algorithm for normal mixture models

Let $\mathbf{Y}_1, \dots, \mathbf{Y}_n$ be p-dimensional random vectors. Assume that an observed data vector $\mathbf{y}_i$ of $\mathbf{Y}_i$ arises from a mixture distribution of G components with density

$$f(\mathbf{y}_i|\boldsymbol{\theta}) = \sum_{k=1}^{G} \lambda_k \phi(\mathbf{y}_i|\boldsymbol{\mu}_k, \Sigma_k), \qquad (1)$$

where $\phi(\mathbf{y}_i|\boldsymbol{\mu}_k, \Sigma_k)$ is the k-th component density of a p-variate normal distribution $N_p(\boldsymbol{\mu}_k, \Sigma_k)$ with mean vector $\boldsymbol{\mu}_k$, variance–covariance matrix Σ_k, λ_k is the k-th mixing proportion such that $0 < \lambda_k < 1$ and $\sum_{k=1}^{G} \lambda_k = 1$, and $\boldsymbol{\theta} = \left[\lambda_1, \dots, \lambda_G, \boldsymbol{\mu}_1^\top, \dots, \boldsymbol{\mu}_G^\top, \mathrm{vec}\Sigma_1^\top, \dots, \mathrm{vec}\Sigma_G^\top\right]^\top$. Here $\mathrm{vec}\Sigma_k$ is the vectorization of Σ_k. The log-likelihood function of $\boldsymbol{\theta}$ for $\mathbf{y} = [\mathbf{y}_1, \dots, \mathbf{y}_n]$ is

$$\ell_o(\boldsymbol{\theta}) = \sum_{i=1}^{n} \log f(\mathbf{y}_i|\boldsymbol{\theta}) = \sum_{i=1}^{n} \log \left\{ \sum_{k=1}^{G} \lambda_k \phi(\mathbf{y}_i|\boldsymbol{\mu}_k, \Sigma_k) \right\}. \qquad (2)$$

Direct maximization of the function (2) is complicated, and then the maximum likelihood estimate (MLE) of $\boldsymbol{\theta}$ is usually found via the EM algorithm [4].

In the setting of the EM algorithm, we regard $\mathbf{y}_i$ as incomplete data and introduce the component-label vector $\mathbf{Z}_i = [Z_{i1}, \dots, Z_{iG}]^\top$ of zero–one indicator variables such that $Z_{ik} = 1$ indicates that $\mathbf{y}_i$ arises from the k-th component of the mixture model and $Z_{ik} = 0$ otherwise. Assume that $\mathbf{Z}_i$ has a multinomial distribution $Mu(1, \boldsymbol{\lambda})$ with parameter $\boldsymbol{\lambda} = [\lambda_1, \dots, \lambda_G]^\top$. In the mixture model, the complete data vector is $\mathbf{x}_i = \left[\mathbf{y}_i^\top, \mathbf{z}_i^\top\right]^\top$, where $\mathbf{y}_i$ is the observed vector and $\mathbf{z}_i$ is the unobserved vector of $\mathbf{Z}_i$. Then $\mathbf{x}_i$ has a mixture distribution with density

$$f(\mathbf{x}_i|\boldsymbol{\theta}) = \prod_{k=1}^{G} \{\lambda_k \phi(\mathbf{y}_i|\boldsymbol{\mu}_k, \Sigma_k)\}^{z_{ik}}. \tag{3}$$

Given $\mathbf{x} = [\mathbf{x}_1, \dots, \mathbf{x}_n]$, the log-likelihood function of $\boldsymbol{\theta}$ is

$$\ell_c(\theta) = \sum_{i=1}^{n} \sum_{k=1}^{G} z_{ik} \log \lambda_k \phi(\mathbf{y}_i|\boldsymbol{\mu}_k, \Sigma_k), \tag{4}$$

and the MLE $\hat{\boldsymbol{\theta}}$ of the function (4) is obtained from

$$\hat{\lambda}_k = \sum_{i=1}^{n} z_{ik}/n, \tag{5}$$

$$\hat{\boldsymbol{\mu}}_k = \sum_{i=1}^{n} z_{ik}\mathbf{x}_i / \sum_{i=1}^{n} z_{ik}, \tag{6}$$

$$\hat{\Sigma}_k = \sum_{i=1}^{n} z_{ik}(\mathbf{x}_i - \hat{\boldsymbol{\mu}}_k)(\mathbf{x}_i - \hat{\boldsymbol{\mu}}_k)^\mathrm{T} / \sum_{i=1}^{n} z_{ik} \tag{7}$$

for $k = 1, \dots, G$. The EM algorithm finds $\hat{\boldsymbol{\theta}}$ by iterating the expectation step (E-step) and the maximization step (M-step). Let $\boldsymbol{\theta}^{(t)}$ be the t-th estimate of $\boldsymbol{\theta}$ in parameter space Θ. The E-step calculates the Q function that is the conditional expectation of $\ell_c(\boldsymbol{\theta})$ given $\mathbf{y}$ and $\boldsymbol{\theta}^{(t)}$ and is written as

$$Q\left(\boldsymbol{\theta}|\boldsymbol{\theta}^{(t)}\right) = E\left[\ell_c(\boldsymbol{\theta})|\mathbf{y}, \boldsymbol{\theta}^{(t)}\right]. \tag{8}$$

Mixture models treat $\mathbf{z} = [\mathbf{z}_1, \dots, \mathbf{z}_n]$ as missing data. The E-step calculates the conditional expectation of Z_{ik} given $\mathbf{y}$ and $\boldsymbol{\theta}^{(t)}$:

$$\begin{aligned} \tau_{ik}^{(t+1)} &= E\left[Z_{ik}|\mathbf{y}, \boldsymbol{\theta}^{(t)}\right] = \Pr\left(Z_{ik}|\mathbf{y}, \boldsymbol{\theta}^{(t)}\right) \\ &= \lambda_k^{(t)} \phi\left(\mathbf{y}_i|\boldsymbol{\mu}_k^{(t)}, \Sigma_k^{(t)}\right) \Big/ \sum_{k=1}^{G} \lambda_k^{(t)} \phi\left(\mathbf{y}_i|\boldsymbol{\mu}_k^{(t)}, \Sigma_k^{(t)}\right). \end{aligned} \tag{9}$$

The quantity $\tau_{ik}^{(t)}$ is the posterior probability that $\mathbf{y}_i$ belongs to the k-th component of the mixture. From Eq. (9), the Q function (8) is given by

$$Q\left(\boldsymbol{\theta}|\boldsymbol{\theta}^{(t)}\right) = \sum_{i=1}^{n} \sum_{k=1}^{G} \tau_{ik}^{(t+1)} \log \lambda_k \phi(\mathbf{y}_i|\boldsymbol{\mu}_k, \Sigma_k). \tag{10}$$

The M-step finds $\boldsymbol{\theta}^{(t+1)}$ maximizing the Q function (10) with respect to $\boldsymbol{\theta}$ over Θ given $\boldsymbol{\theta}^{(t)}$:

$$\boldsymbol{\theta}^{(t+1)} = \arg\max_{\boldsymbol{\theta} \in \Theta} Q\left(\boldsymbol{\theta}|\boldsymbol{\theta}^{(t)}\right). \tag{11}$$

When replacing z_{ik} in Eq. (5) with $\tau_{ik}^{(t+1)}$ in the E-step, we obtain

$$\lambda_k^{(t+1)} = \frac{1}{n}\sum_{i=1}^{n} \tau_{ik}^{(t+1)}. \tag{12}$$

From Eqs. (6) and (7), we also have

$$\boldsymbol{\mu}_k^{(t+1)} = \sum_{i=1}^{n} \tau_{ik}^{(t+1)} \mathbf{x}_i \Big/ \sum_{i=1}^{n} \tau_{ik}^{(t+1)}, \tag{13}$$

$$\hat{\Sigma}_k^{(t+1)} = \sum_{i=1}^{n} \tau_{ik}^{(t+1)} \left(\mathbf{x}_i - \boldsymbol{\mu}_k^{(t+1)}\right)\left(\mathbf{x}_i - \boldsymbol{\mu}_k^{(t+1)}\right)^{\mathrm{T}} \Big/ \sum_{i=1}^{n} \tau_{ik}^{(t+1)}. \tag{14}$$

We describe the EM algorithm for the normal mixture model in Algorithm 1.

Algorithm 1: The EM algorithm.

E-step: Calculate $\boldsymbol{\tau}_k^{(t+1)} = \left[\tau_{i1}^{(t+1)}, \dots, \tau_{iG}^{(t+1)}\right]^{\mathrm{T}}$ using Eq. (9) and update $\boldsymbol{\tau}^{(t+1)} = \left[\boldsymbol{\tau}_1{}^{(t+1)}, \dots, \boldsymbol{\tau}_n^{(t+1)}\right]$.

M-step: Estimate $\boldsymbol{\theta}^{(t+1)}$ from Eqs. (12)–(14).

---

3. Acceleration of the EM algorithm

In order to accelerate the convergence of the EM algorithm, we can use the ε-accelerated EM algorithm [9] and the εR-accelerated EM algorithm [13]. The ε-accelerated EM algorithm incorporates the vector ε (vε) algorithm [10] in the EM algorithm. The εR-accelerated EM algorithm improves the computation speed of the ε-accelerated EM algorithm by adding a restarting procedure.

We briefly introduce the vε algorithm. Let $\left\{\boldsymbol{\theta}^{(t)}\right\}_{t\geq 0}$ be a linearly convergent vector sequence from an iterative computational procedure and converge to a stationary point $\hat{\boldsymbol{\theta}}$ as $t \to \infty$. Then the vε algorithm generates a sequence $\left\{\boldsymbol{\psi}^{(t)}\right\}_{t\geq 0}$ that converges to $\hat{\boldsymbol{\theta}}$ faster than $\left\{\theta^{(t)}\right\}_{t\geq 0}$ by using

$$\psi^{(t-1)} = \boldsymbol{\theta}^{(t)} + \left[\left[\Delta\boldsymbol{\theta}^{(t)}\right]^{-1} - \left[\Delta\boldsymbol{\theta}^{(t-1)}\right]^{-1}\right]^{-1}, \tag{15}$$

where $\Delta\boldsymbol{\theta}^{(t)} = \boldsymbol{\theta}^{(t+1)} - \boldsymbol{\theta}^{(t)}$ and $[\boldsymbol{\theta}]^{-1} = \boldsymbol{\theta}/\|\boldsymbol{\theta}\|^2 = \boldsymbol{\theta}/\boldsymbol{\theta}^{\mathrm{T}}\boldsymbol{\theta}$, see Appendix A for details. The algorithm enables accelerating the convergence of a slowly convergent vector sequence and is very effective for linearly convergent sequences.

We define the EM algorithm as a mapping $\boldsymbol{\theta} \mapsto M(\boldsymbol{\theta})$ from Θ to Θ such that each iteration $\boldsymbol{\theta}^{(t)} \to \boldsymbol{\theta}^{(t+1)}$ is denoted by

$$\boldsymbol{\theta}^{(t+1)} = M\left(\boldsymbol{\theta}^{(t)}\right). \tag{16}$$

Algorithm 2: The ε-accelerated EM algorithm.

E-step: Estimate $\boldsymbol{\theta}^{(t+1)}$ from Eq. (16).
ε **acceleration step** Calculate $\psi^{(t-1)}$ from $\left\{\boldsymbol{\theta}^{(t+1)}, \boldsymbol{\theta}^{(t)}, \boldsymbol{\theta}^{(t-1)}\right\}$ using Eq. (15).

The ε-accelerated EM algorithm is shown in Algorithm 2. Given a convergence criterion δ, the ε-accelerated EM algorithm iterates until

$$\|\psi^{(t-1)} - \psi^{(t-2)}\|^2 < \delta. \tag{17}$$

Assume that the sequence $\left\{\boldsymbol{\theta}^{(t)}\right\}_{t\geq 0}$ from the EM algorithm converges to a stationary point $\hat{\boldsymbol{\theta}}$. The εR-accelerated EM algorithm generates $\{\psi^{(t)}\}_{t\geq 0}$ converging to $\hat{\boldsymbol{\theta}}$ faster than $\left\{\boldsymbol{\theta}^{(t)}\right\}_{t\geq 0}$ and provides $\hat{\boldsymbol{\theta}}$ from the final value of $\{\psi^{(t)}\}_{t\geq 0}$ when the algorithm terminates.

The theorems with the convergence and acceleration of the algorithm are given in [18].

As shown in Algorithm 2, the ε-accelerated EM algorithm generates two parallel sequences, $\{\psi^{(t)}\}_{t\geq 0}$ in the ε acceleration step and $\left\{\boldsymbol{\theta}^{(t)}\right\}_{t\geq 0}$ in the EM step. At the ε acceleration step, the EM estimate $M(\psi^{(t-1)})$ from $\psi^{(t-1)}$ may have a larger log-likelihood function than the current EM estimate $\boldsymbol{\theta}^{(t+1)}$, that is,

$$\ell_o\left(M\left(\psi^{(t-1)}\right)\right) > \ell_o\left(\boldsymbol{\theta}^{(t+1)}\right). \tag{18}$$

When this occurs, the EM step is restarted with $M(\psi^{(t-1)})$ as the initial value, and the ε acceleration step gets $\psi^{(t)}$ from $\{\psi^{(t-1)}, M(\psi^{(t-1)}), M(M(\psi^{(t-1)}))\}$. Notice that at the restarting point, we still generate the EM sequence using three estimates obtained from the same initial value $\psi^{(t-1)}$. By this manner, we keep to always apply the ε-acceleration to a sequence obtained by the EM mapping M from the same initial value.

By our experiments, the restarting procedure is performed almost every time when we only use the restarting condition $\ell_o(M(\psi^{(t-1)})) > \ell_o\left(\boldsymbol{\theta}^{(t+1)}\right)$, and then it inefficiently takes much computation time. As one more condition for restarting the EM step, we give $\|\psi^{(t-1)} - \psi^{(t-2)}\|^2 \leq \delta_{Re}(\ > \delta)$ and reset $\delta_{Re} = \delta_{Re}/10^k$ at each restarting, where k is an integer, such as one. By adding this condition, we can control the restarting frequency. For example, set $\delta = 10^{-12}$, and initialize $\delta_{Re} = 1$ and $k = 1$. Then the restarting procedure is performed at most 12 times.

The restarting conditions are summarized as follows:

i. $\ell_o(M(\psi^{(t-1)})) > \ell_o\left(\boldsymbol{\theta}^{(t+1)}\right)$, and

ii. $\|\psi^{(t-1)} - \psi^{(t-2)}\|^2 < \delta_{Re}$.

Condition (i) means that the log-likelihood function can be increased by restarting. Condition (ii) is used to reduce the frequency of restarting. This is the key idea of the restarting procedure. The εR-accelerated EM algorithm is the ε-accelerated EM algorithm with the restarting procedure using conditions (i) and (ii) and is given in Algorithm 3.

Algorithm 3: The εR-accelerated EM algorithm.

EM step: Estimate $\boldsymbol{\theta}^{(t+1)}$ from Eq. (16).

ε **acceleration step:** Calculate $\psi^{(t-1)}$ from $\left\{\boldsymbol{\theta}^{(t+1)}, \boldsymbol{\theta}^{(t)}, \boldsymbol{\theta}^{(t-1)}\right\}$ using Eq. (15).

Restarting step: If $\ell_o\left(M\left(\psi^{(t-1)}\right)\right) > \ell_o\left(\boldsymbol{\theta}^{(t+1)}\right)$ and $\|\psi^{(t-1)} - \psi^{(t-2)}\|^2 < \delta_{Re}$, then set

$$\boldsymbol{\theta}^{(t)} = \psi^{(t-1)}, \tag{19}$$

update

$$\boldsymbol{\theta}^{(t+1)} = M\left(\psi^{(t-1)}\right), \tag{20}$$

and reset

$$\delta_{Re} = \delta_{Re}/10^k. \tag{21}$$

The εR-accelerated EM algorithm also gives $\hat{\boldsymbol{\theta}}$ from the final value of $\left\{\psi^{(t)}\right\}_{t\geq 0}$. When the restarting step effectively finds values for restating the EM step, the εR-accelerated EM algorithm greatly reduces the number of iterations and computation time for convergence. The advantage of the εR-accelerated EM algorithm over the ε-accelerated EM algorithm is that it restarts the EM step at a better current estimate and also keeps that the log-likelihood function increases in the iterations.

Theoretical results of convergence and speed of convergence of the εR-accelerated EM algorithm are given in [13].

4. Numerical experiments for the acceleration of the EM algorithm

We investigate how much faster the ε-accelerated EM and εR-accelerated EM algorithms converge than the EM algorithm. All computations are performed with the statistical package R [19] executing on Windows, Intel Core i5 3.00 GHz with 8 GB of memory.

The R package `MixSim` [17, 20] is used to simulate a random data matrix $\mathbf{y}$ having a p-variate normal mixture distribution of G components. We generate $\mathbf{y} = \left[\mathbf{y}_1, \dots, \mathbf{y}_{1000}\right]$ and find the MLE of $\boldsymbol{\theta}$ using the EM, ε-accelerated EM, and εR-accelerated EM algorithms. The procedure is replicated 100 times. Here, we consider $p = 2, 3, 4, 5, 6$ and $G = 4$. For all experiments, we set $\delta = 10^{-12}$ for convergence of the algorithms, $\delta_{Re} = 1$ and $k = 1$ for the restarting condition of the εR-accelerated EM algorithm. Initial values of the algorithms are obtained from the k-means method using the R function `kmeans`.

Tables 1 and **2** report the results of the number of iterations and CPU time of these algorithms for each p. The CPU times (in seconds) are measured by the R function `proc.time` that times are typically available to 10 milliseconds. For all computations, the acceleration algorithms found the same MLEs as those from the EM algorithm. We see from the tables that the EM algorithm requires a large number of iterations for convergence, whereas two acceleration algorithms converge a smaller number of iterations than the EM algorithm. Then the εR-accelerated EM algorithm can greatly reduce both the number of iterations and CPU time.

To measure the speed of convergence of the EM and two acceleration algorithms, we calculate iteration and CPU time speedups. The iteration speedup of an acceleration algorithm for the EM algorithm is defined by

$$\frac{\text{The number of iterations of the EM algorithm}}{\text{The number of iterations of an acceleration algorithm}}.$$

		Min.	1st Qu.	Median	Mean	3rd Qu.	Max.
$p = 2$	EM	172.00	467.25	771.00	1069.48	1302.25	10852.00
	ε	133.00	308.50	445.00	697.74	706.50	8090.00
	εR	83.00	182.50	253.50	424.22	396.50	4967.00
$p = 3$	EM	210.00	403.50	628.50	716.33	946.75	1973.00
	ε	121.00	276.75	400.50	484.83	604.75	1566.00
	εR	68.00	167.50	244.50	307.99	359.75	1183.00
$p = 4$	EM	166.00	372.75	468.50	618.63	755.75	2193.00
	ε	120.00	248.75	331.50	400.00	461.50	1452.00
	εR	58.00	139.00	194.50	241.25	291.25	884.00
$p = 5$	EM	141.00	334.75	492.50	879.35	783.00	24886.00
	ε	101.00	235.50	351.00	687.31	516.00	24756.00
	εR	57.00	144.00	226.00	431.55	336.50	14288.00
$p = 6$	EM	193.00	361.25	499.00	655.80	647.75	5910.00
	ε	144.00	252.00	323.50	454.45	473.75	5825.00
	εR	99.00	163.75	230.50	302.13	299.00	4771.00

Table 1.
Summary statistics of the number of iterations of the EM, ε-accelerated EM (ε) and εR-accelerated EM (εR) algorithms for 100 simulated random data. Each data is generated from a p-variate normal mixture distribution of four components.

		Min.	1st Qu.	Median	Mean	3rd Qu.	Max.
$p = 2$	EM	0.39	1.04	1.68	2.31	2.80	22.73
	ε	0.30	0.75	1.08	1.66	1.66	19.18
	εR	0.22	0.49	0.66	1.11	1.04	13.21
$p = 3$	EM	0.75	1.40	2.07	2.64	3.30	8.53
	ε	0.45	1.01	1.46	1.99	2.52	7.60
	εR	0.35	0.68	1.00	1.44	1.68	8.26
$p = 4$	EM	0.42	0.93	1.16	1.53	1.86	5.34
	ε	0.28	0.65	0.86	1.06	1.24	3.80
	εR	0.20	0.44	0.59	0.71	0.86	2.39
$p = 5$	EM	0.25	0.64	0.92	1.65	1.50	46.11
	ε	0.22	0.49	0.72	1.42	1.08	50.36
	εR	0.16	0.35	0.51	0.95	0.80	29.07
$p = 6$	EM	0.51	1.02	1.42	1.84	1.88	17.86
	ε	0.43	0.75	1.02	1.37	1.47	17.75
	εR	0.32	0.54	0.76	0.99	1.00	14.29

Table 2.
Summary statistics of CPU time of the EM, ε-accelerated EM (ε) and εR-accelerated EM (εR) algorithms for 100 random data. Each data is generated from a p-variate normal mixture distribution of four components.

The CPU time speedup is also calculated similarly to the iteration speedup. **Tables 3** and **4** show the results of the iteration and CPU time speedups of two acceleration algorithms. We compare the mean values of the iteration and CPU time speedups of the algorithms. The ε-accelerated EM algorithm is about 1.5 times and 1.4 times faster than the EM algorithm in the number of iterations and CPU time, respectively. Then the εR-accelerated EM algorithm is more than twice as fast as the EM algorithm in both the number of iterations and CPU time. The boxplots of **Figures 1** and **2** also show that the εR-accelerated EM algorithm is obviously much faster than the ε-accelerated EM algorithm. **Table 3** and **Figure 1** indicate that in 75 out of 100 replications, the number of iterations of the εR-accelerated EM algorithm is less than half as small as that of the EM algorithm. For CPU time of **Table 4** and **Figure 2**, the εR-accelerated EM algorithm is more than twice as fast as the EM algorithm in 50 out of 100 replications.

Figure 3 shows the boxplots of the iteration and CPU time speedups of the εR-accelerated EM algorithm for $p = 6$. Here, "more" ("less") means that the number of iterations of the EM algorithm is more (less) than the median in **Tables 1** and **2**.

		Min.	**1st Qu.**	**Median**	**Mean**	**3rd Qu.**	**Max.**
$p = 2$	ε	1.05	1.34	1.54	1.61	1.77	3.58
	εR	1.15	2.08	2.73	3.03	3.48	11.36
$p = 3$	ε	1.07	1.32	1.52	1.52	1.68	2.15
	εR	1.20	1.97	2.57	2.58	2.98	6.08
$p = 4$	ε	1.13	1.32	1.48	1.51	1.62	2.33
	εR	1.45	2.09	2.42	2.60	2.94	9.04
$p = 5$	ε	1.01	1.30	1.46	1.47	1.63	2.06
	εR	1.33	1.84	2.23	2.32	2.67	4.32
$p = 6$	ε	1.01	1.28	1.46	1.49	1.65	2.33
	εR	1.24	1.86	2.17	2.37	2.59	6.75

Table 3.
Summary statistics of the iteration speedup of the ε-accelerated EM (ε) and εR-accelerated EM (εR) algorithms for 100 random data. Each data is generated from a p-variate normal mixture distribution of four components.

		Min.	**1st Qu.**	**Median**	**Mean**	**3rd Qu.**	**Max.**
$p = 2$	ε	0.97	1.22	1.45	1.47	1.67	3.37
	εR	1.05	1.71	2.24	2.50	2.85	8.60
$p = 3$	ε	0.85	1.21	1.39	1.40	1.56	2.07
	εR	0.78	1.61	2.04	2.08	2.40	4.48
$p = 4$	ε	1.02	1.27	1.39	1.43	1.53	2.11
	εR	1.20	1.70	2.03	2.17	2.43	7.48
$p = 5$	ε	0.92	1.17	1.33	1.34	1.50	2.06
	εR	1.12	1.48	1.76	1.86	2.12	3.21
$p = 6$	ε	0.84	1.18	1.39	1.39	1.55	2.21
	εR	1.00	1.57	1.77	1.98	2.24	5.47

Table 4.
Summary statistics of the CPU time speedup of the ε-accelerated EM (ε) and εR-accelerated EM (εR) algorithms for 100 random data. Each data is generated from p-variate normal mixture distributions of four components.

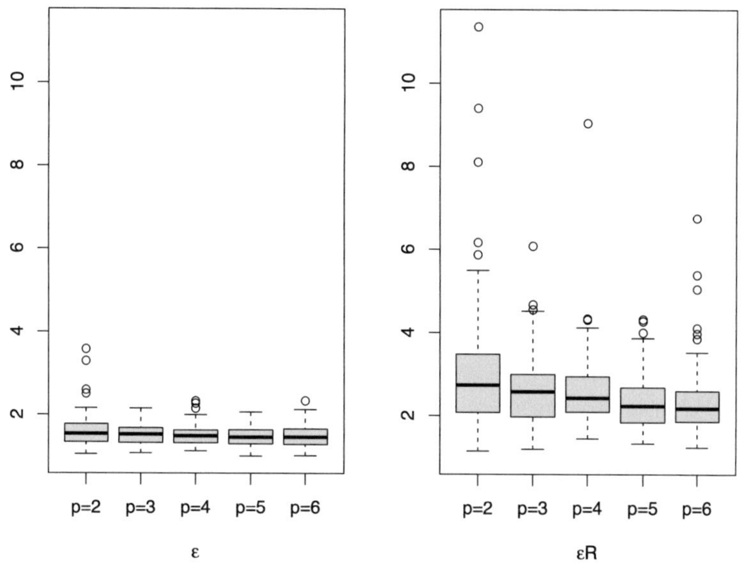

Figure 1.
Boxplots of the iteration speedup of the ε-accelerated EM (ε) and εR-accelerated EM (εR) algorithms for 100 random data generated from a p-variate normal mixture distribution of four components.

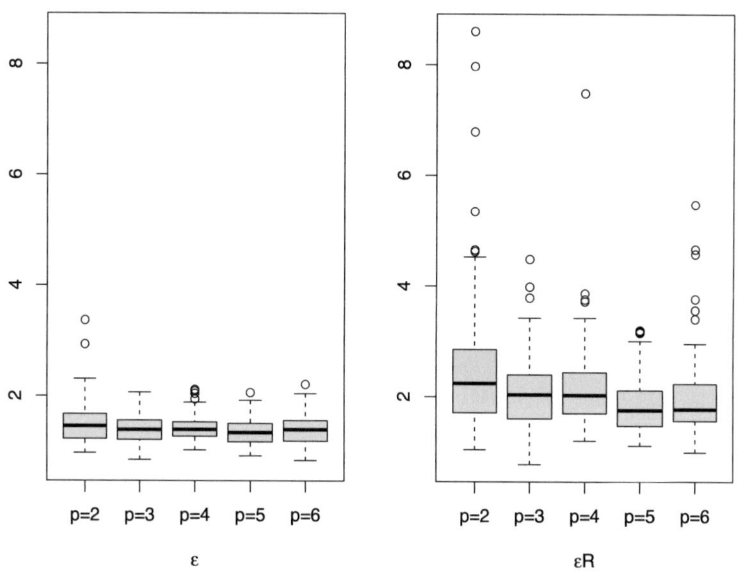

Figure 2.
Boxplots of the CPU time speedup of the ε-accelerated EM (ε) and εR-accelerated EM (εR) algorithms for 100 random data. Each data is generated from a p-variate normal mixture distribution of four components.

We can see from the figure that, for the larger number of iterations of the EM algorithm ("more"), the εR-accelerated EM algorithm works well to speed up the convergence of $\{\psi^{(t)}\}_{t\geq 0}$. We observed a similar result for other p. Therefore, the algorithm is more powerful when the EM algorithm takes a larger number of iterations.

The results from the tables and figures demonstrate that the restarting step in the εR-accelerated EM algorithm enables a significant increase in the computation speed with less computational effort.

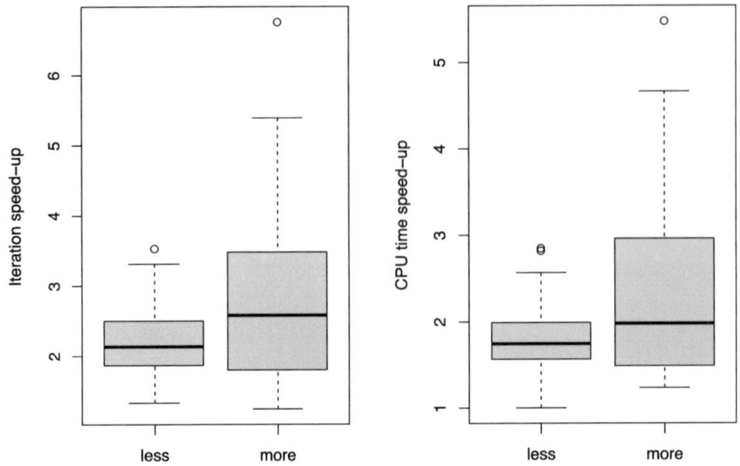

Figure 3.
Boxplots of the iteration and CPU time speedups of the εR-accelerated EM algorithms for 100 random data. Each data is generated from a six-variate normal mixture distribution of four components. The label "less" ("more") means that the number of iterations of the EM algorithm is less (more) than the median in ***Tables 1*** *and 2.*

5. Initial value selection for normal mixture models

It is well known that the log-likelihood function (2) may have numerous maximums. The EM algorithm does not guarantee to obtain the global maximum of the log-likelihood function due to its local convergence. Thus, the initial value of $\boldsymbol{\theta}$ deeply depends on the performance of the EM algorithm. Several methods for selecting the initial value are proposed; for example, see [14–17]. These methods are based on the multiple runs of the EM algorithm using different initial values and find $\hat{\boldsymbol{\theta}}$ for getting the global maximum of the log-likelihood function.

We apply the emEM algorithm [14] to the mixture model estimation. The algorithm is a popular one and usually provides excellent results when the number of components is not large [21]. The emEM algorithm selects an initial value in the em step that is several short runs of the EM algorithm using different initial values and a lax convergence criterion and obtains $\hat{\boldsymbol{\theta}}$ from the EM step that runs the EM algorithm starting from the initial value with a strict convergence criterion.

The em step consists of three steps. The first step generates J initial values of $\boldsymbol{\theta}$. The second step runs the EM algorithm from these initial values with a lax convergence criterion. Hence, we do not wait for convergence of the EM algorithm and stop the iterations. The third step selects the value giving the largest log-likelihood function among J trials.

Let δ_{ini} be a convergence criterion and T_{max} the maximum number of iterations. We present the emEM algorithm in Algorithm 4.

Algorithm 4: The emEM algorithm.

---

em step: Select $\boldsymbol{\theta}^{(0)}$ of the EM step.

Random initialization step: Draw J initial values $\left\{\boldsymbol{\theta}^{(0,j)}\right\}_{j=1,\dots,J}$.

Short running step: Repeat the following computation for $j = 1, \dots, J$:
Generate $\left\{\boldsymbol{\theta}^{(t_j,j)}\right\}_{t_j \geq 0}$ by iterating the EM algorithm from $\boldsymbol{\theta}^{(0,j)}$ and stop the iterations at the t_j-iteration if

---

$$\frac{\ell_o\left(\boldsymbol{\theta}^{(t_j,j)}\right) - \ell_o\left(\boldsymbol{\theta}^{(t_j-1,j)}\right)}{\ell_o\left(\boldsymbol{\theta}^{(t_j,j)}\right) - \ell_o\left(\boldsymbol{\theta}^{(0,j)}\right)} < \delta_{ini}, \quad \text{or} \quad t_j > T_{max}. \tag{22}$$

Obtain $\boldsymbol{\theta}^{(*,j)} = \boldsymbol{\theta}^{(t_j,j)}$.

Selection step: From J candidate initial values $\left\{\boldsymbol{\theta}^{(*,j)}\right\}_{j=1,\dots,J}$, find

$$\boldsymbol{\theta}^{(0)} = \underset{\left\{\boldsymbol{\theta}^{(*,j)}\right\}_{j=1,\dots,J}}{\arg\max} \left\{\ell_o\left(\boldsymbol{\theta}^{(*,j)}\right)\right\}_{j=1,\dots,J}. \tag{23}$$

EM step: Given $\boldsymbol{\theta}^{(0)}$ in the em step, find $\hat{\boldsymbol{\theta}}$ using the EM algorithm.

The em step performs multiple runs of the EM algorithm, and then its computation may be time-consuming. We replace the EM algorithm with the ε-accelerated EM algorithm in the em step and use the εR-accelerated EM algorithm to obtain $\hat{\theta}$ in the EM step. By applying these acceleration algorithms to the emEM algorithm, it is possible to reduce the number of iterations and CPU time. The acceleration of the emEM algorithm is referred as to the εem-εREM algorithm and is shown in Algorithm 5.

Algorithm 5: the εem-εREM algorithm.

ε-**em step**: Select $\boldsymbol{\theta}^{(0)}$ of the εR-EM step.

Random initialization step: Draw J initial values $\left\{\boldsymbol{\theta}^{(0,j)}\right\}_{j=1,\dots,J}$.

Short running step: Repeat the following computation for $j = 1, \dots, J$:

Generate $\left\{\psi^{(t_j,j)}\right\}_{t_j \geq 0}$ by iterating the ε-accelerated EM algorithm from $\boldsymbol{\theta}^{(0,j)}$ and stop the iterations at the t_j-iteration if

$$\frac{\ell_o\left(\psi^{(t_j,j)}\right) - \ell_o\left(\psi^{(t_j-1,j)}\right)}{\ell_o\left(\psi^{(t_j,j)}\right) - \ell_o\left(\psi^{(0,j)}\right)} < \delta_{ini}, \quad \text{or} \quad t_j > T_{max}. \tag{24}$$

Obtain $\boldsymbol{\theta}^{(*,j)} = \psi^{(t_j,j)}$.

Selection step: From J candidate initial values $\left\{\boldsymbol{\theta}^{(*,j)}\right\}_{j=1,\dots,J}$, find

$$\boldsymbol{\theta}^{(0)} = \underset{\left\{\boldsymbol{\theta}^{(*,j)}\right\}_{j=1,\dots,J}}{\arg\max} \left\{\ell_o\left(\boldsymbol{\theta}^{(*,j)}\right)\right\}_{j=1,\dots,J}. \tag{25}$$

ε-**R-EM step**: Given $\boldsymbol{\theta}^{(0)}$ in the em step, find $\hat{\theta}$ using the εR-accelerated EM algorithm.

---

6. Numerical experiments for the initial value selection

We evaluate the performance of the ε-accelerated EM and εR-accelerated EM algorithms in application to the emEM algorithm.

	emEM			εem-εREM		
	em	EM	total	ε-em	εR-EM	total
$p = 2$	1912	3834	5746	1415	1429	2844
$p = 3$	1995	1490	3485	925	354	1279
$p = 4$	2352	725	3077	997	451	1448
$p = 5$	3344	885	4229	1516	397	1913
$p = 6$	2641	957	3598	1234	435	1669

Table 5.
The numbers of iterations of the emEM and εem-εREM algorithms. The em and ε-em steps generate 50 random initial values.

	emEM			εem-εREM		
	em	EM	total	ε-em	εR-EM	total
$p = 2$	6.04	7.37	13.41	4.67	3.22	7.89
$p = 3$	6.36	3.14	9.50	3.23	1.00	4.23
$p = 4$	8.81	1.61	10.42	3.98	1.86	5.84
$p = 5$	12.55	2.33	14.88	6.04	1.19	7.23
$p = 6$	11.01	2.44	13.45	5.35	1.43	6.78

Table 6.
CPU times of the emEM and εem-εREM algorithms. The em and ε-em steps generate 50 random initial values.

By using MixSim, we simulate $\mathbf{y} = [\mathbf{y}_1, \dots, \mathbf{y}_{1000}]$ having the p-variate normal mixture distribution of six components for $p = 2, 3, 4, 5, 6$. The values of δ, δ_{Re}, and k are the same as in the experiments of Section 1.4. Assume that the probability of not finding the global maximum of the log-likelihood function in a single run is 0.80 for safety. Then the probability of finding the global maximum at least once is $1 - 0.80^{50} > 0.9999$. In the em and ε-em steps, we draw 50 initial values $\left\{\boldsymbol{\theta}^{(0,j)}\right\}_{j=1,\dots,50}$ from kmeans and set $\delta_{ini} = 0.001$ and $T_{max} = 1000$.

Tables 5 and **6** present the number of iterations and CPU time for each p. We see from **Table 5** that the number of iterations of the ε-em step is much smaller than that of the em step. The ε-accelerated EM algorithm effectively improves the computation speed of the em step. We compare the number of iterations and CPU time of the εem-εREM algorithm with those of the emEM algorithm. Then these values of the εem-εREM algorithm are about less than half of those of the emEM algorithm. The results illustrate that the ε-accelerated EM and εR-accelerated EM algorithms can sufficiently accelerate the convergence of the emEM algorithm.

7. Concluding remarks

In this chapter, we introduced the ε-accelerated EM and εR-accelerated EM algorithms. Both algorithms are given by very simple computational procedures and are executed with a little bit of computation for each iteration, while they well accelerate the convergence of the EM algorithm.

When the EM algorithm is applied to normal mixture models, the algorithm may converge slowly and be heavily dependent on the initial value. The first problem is solved by the acceleration of the EM algorithm. The numerical experiments indicated the availability of the ε-accelerated EM and εR-accelerated EM algorithms. For the second problem, the initial value selection is useful to initiate the EM algorithm. We applied the emEM algorithm to normal mixture model estimation and developed the εem-εREM algorithm to speed up the computation of the emEM algorithm. Then the ε-accelerated EM algorithm is used in the em step, and the εR-accelerated EM algorithm is in the EM step. Numerical experiments showed that the εem-εREM algorithm can converge in a smaller number of iterations and shorter CPU time than the emEM algorithm.

The ε-accelerated EM and εR-accelerated EM algorithms accelerate the convergence of the EM algorithm without any modification of the E- and M-steps of the algorithm. This means that these algorithms do not require to derive the acceleration formula for every statistical model. Thus, these algorithms are applied to several mixture models—mixtures of factor analyzers, mixtures of multivariate t-distributions, mixtures of generalized hyperbolic distributions, and parsimonious Gaussian mixture models. We expect that the convergence of the EM algorithms used in these mixture models tends to be slow. The results from the experiments show that the εR-accelerated EM and εR-accelerated EM algorithms are useful due to their fast speed of convergence and ease of use.

Appendix: the vector ε algorithm

Let $\boldsymbol{\theta}^{(t)}$ denote a d-dimensional vector that converges to a vector $\hat{\boldsymbol{\theta}}$ as $t \to \infty$. We define $[\boldsymbol{\theta}]^{-1} = \boldsymbol{\theta}/\|\boldsymbol{\theta}\|^2 = \boldsymbol{\theta}/\boldsymbol{\theta}^\top\boldsymbol{\theta}$. In general, the v$\varepsilon$ algorithm for a sequence $\left\{\boldsymbol{\theta}^{(t)}\right\}_{t\geq 0}$ starts with

$$\varepsilon^{(t,-1)} = \mathbf{0}, \qquad \varepsilon^{(t,0)} = \boldsymbol{\theta}^{(t)} \tag{26}$$

and then generates a vector $\varepsilon^{(t,k+1)}$ by

$$\begin{aligned}\varepsilon^{(t,k+1)} &= \varepsilon^{(t+1,k-1)} + \left[\varepsilon^{(t+1,k)} - \varepsilon^{(t,k)}\right] \\ &= \varepsilon^{(t+1,k-1)} + \left[\Delta\varepsilon^{(t,k)}\right]^{-1}, \qquad k = 0, 1, 2, \ldots .\end{aligned} \tag{27}$$

For practical implementation, we apply the vε algorithm for $k = 1$ to accelerate the convergence of $\left\{\boldsymbol{\theta}^{(t)}\right\}_{t\geq 0}$. From the above equation, we have

$$\varepsilon^{(t,2)} = \varepsilon^{(t+1,0)} + \left[\Delta\varepsilon^{(t,1)}\right]^{-1} \quad \text{for } k = 1, \tag{28}$$

$$\varepsilon^{(t,1)} = \varepsilon^{(t+1,-1)} + \left[\Delta\varepsilon^{(t,0)}\right]^{-1} = \left[\Delta\varepsilon^{(t,0)}\right]^{-1} \quad \text{for } k = 0. \tag{29}$$

Then the vector $\varepsilon^{(t,2)}$ becomes as follows:

$$\begin{aligned}\varepsilon^{(t,2)} &= \varepsilon^{(t+1,0)} + \left[\left[\Delta\varepsilon^{(t+1,0)}\right]^{-1} - \left[\Delta\varepsilon^{(t,0)}\right]^{-1}\right]^{-1} \\ &= \boldsymbol{\theta}^{(t+1)} + \left[\left[\Delta\boldsymbol{\theta}^{(t+1)}\right]^{-1} - \left[\Delta\boldsymbol{\theta}^{(t)}\right]^{-1}\right]^{-1}.\end{aligned} \tag{30}$$

When setting $\psi^{(t)} = \varepsilon^{(t,2)}$, we obtain Eq. (15).

Author details

Masahiro Kuroda
Okayama University of Science, Okayama City, Japan

*Address all correspondence to: kuroda@mgt.ous.ac.jp

References

[1] Bouveyron C, Celeux G, Murphy TB, Raftery AE. Model-Based Clustering and Classification for Data Science with Applications in R. Cambridge: Cambridge University Press; 2019

[2] McLachlan G, Peel D. Finite Mixture Models. New York: Wiley; 2000

[3] McNicholas PD. Mixture Model-Based Classification. Boca Raton Chapman & Hall/CRC Press; 2016

[4] Dempster AP, Laird NM, Rubin DB. Maximum likelihood from incomplete data via the EM algorithm. With discussion. Journal of the Royal Statistical Society Series B. 1977;**39**:1-38

[5] Louis TA. Finding the observed information matrix when using the EM algorithm. Journal of the Royal Statistical Society, Series B. 1982;**44**: 226-233

[6] Jamshidian M, Jennrich RI. Conjugate gradient acceleration of the EM algorithm. Journal of the American Statistical Association. 1993; **88**:221-228

[7] Jamshidian M, Jennrich RI. Acceleration of the EM algorithm by using quasi-Newton methods. Journal of the Royal Statistical Society, Series B. 1997;**59**:569-587

[8] Lange K. A quasi Newton acceleration of the EM algorithm. Statistica Sinica. 1995;**5**:1-18

[9] Kuroda M, Sakakihara M. Accelerating the convergence of the EM algorithm using the vector ε algorithm. Computational Statistics & Data Analysis. 2006;**51**:1549-1561

[10] Wynn P. Acceleration techniques for iterated vector and matrix problems. Mathematics of Computation. 1962;**16**: 301-322

[11] Brezinski C, Redivo-Zaglia M. Extrapolation Methods: Theory and Practice. Amsterdam: North-Holland; 1991

[12] Smith DA, Ford F, Sidi A. Extrapolation methods for vector sequences. SIAM Review. 1987;**29**: 199-233

[13] Kuroda M, Geng Z, Sakakihara M. Improving the vector ε acceleration for the EM algorithm using a re-starting procedure. Computational Statistics. 2015;**30**:1051-1077

[14] Biernacki C, Celeux G, Govaert G. Choosing starting values for the EM algorithm for getting the highest likelihood in multivariate Gaussian mixture models. Computational Statistics & Data Analysis. 2003;**41**: 561-575

[15] Kwedlo W. A new random approach for initialization of the multiple restart EM algorithm for Gaussian model-based clustering. Pattern Analysis and Applications. 2015;**18**:757-770

[16] Maitra R. Initializing optimization partitioning algorithms. IEEE/ACM Transactions on Computational Biology and Bioinformatics. 2009;**6**: 144-157

[17] Melnykov V, Chen W, Maitra R. MixSim: An R package for simulating data to study performance of clustering algorithms. Journal of Statistical Software. 2012;**51**:1

[18] Wang M, Kuroda M, Sakakihara M, Geng Z. Acceleration of the EM algorithm using the vector epsilon algorithm. Computational Statistics. 2008;**23**:469-486

[19] R Core Team. R. A Language and Environment for Statistical Computing.

Vienna, Austria: R Foundation for Statistical Computing; 2021; Available from: https://www.R-project.org/

[20] Maitra R, Melnykov V. Simulating data to study performance of finite mixture modeling and clustering algorithms. Journal of Computational and Graphical Statistics. 2010;**19**: 354-376

[21] Michael S, Melnykov V. An effective strategy for initializing the EM algorithm in finite mixture models. Advances in Data Analysis and Classification. 2016;**10**:563-583

Section 2

Frailty Models

Chapter 5

Dependent Dirichlet Processes for Analysis of a Generalized Shared Frailty Model

Chong Zhong, Zhihua Ma, Junshan Shen and Catherine Liu

Abstract

Bayesian paradigm takes advantage of well-fitting complicated survival models and feasible computing in survival analysis owing to the superiority in tackling the complex censoring scheme, compared with the frequentist paradigm. In this chapter, we aim to display the latest tendency in Bayesian computing, in the sense of automating the posterior sampling, through a Bayesian analysis of survival modeling for multivariate survival outcomes with the complicated data structure. Motivated by relaxing the strong assumption of proportionality and the restriction of a common baseline population, we propose a generalized shared frailty model which includes both parametric and nonparametric frailty random effects to incorporate both treatment-wise and temporal variation for multiple events. We develop a survival-function version of the ANOVA dependent Dirichlet process to model the dependency among the baseline survival functions. The posterior sampling is implemented by the No-U-Turn sampler in Stan, a contemporary Bayesian computing tool, automatically. The proposed model is validated by analysis of the bladder cancer recurrences data. The estimation is consistent with existing results. Our model and Bayesian inference provide evidence that the Bayesian paradigm fosters complex modeling and feasible computing in survival analysis, and Stan relaxes the posterior inference.

Keywords: ANOVA DDP, dependent treatments, multivariate survival outcomes, recurrence, Stan

1. Introduction

The shared frailty model, coined by [1], has been widely used in the analysis of multivariate survival outcomes that might be associated with subgroups or clusters. Enormous work has been devoted to the development of the shared frailty model in both Bayesian and frequency paradigms, and the reviews can be found in [2, 3]. As an extension of the well-known Cox's proportional hazard model, *conditional on the frailty effect*, the traditional shared frailty model assumes a proportional hazards structure, that is, the hazard ratio between two sets of covariate values is proportional to their difference in relative risk scores over time [4]. Meanwhile, it fixes the baseline hazard function among all clusters.

Traditional shared frailty models provide a good framework for expediently mathematical tractability of the heterogeneity among multivariate observations, whereas in practice it needs modification and adaption to tolerate complex structure so as to incorporate cross information owing to the intra- and inter-subject variability [5, 6]. Take the renowned data on recurrences of bladder cancer, for instance [7]. There are three treatment arms, placebo, thiotepa, and pyridoxine. Patients had multiple recurrences of tumors which were sparse beyond the fourth recurrence. **Figure 1** displays the Kaplan-Meier estimators of the survival-function for the times of the first and the second recurrences under three treatments. One observes that in **Figure 1(a)**, the estimated survival curves at the first recurrence are crossed, indicating a crossed hazard, and therefore, the proportional hazard assumption is suspected [8]; in **Figure 1(b)**, the survival curve of pyridoxine drops below that of placebo from the fifth month at the second recurrence rather than above from the 10th month on at the first recurrence. This indicates the functional form of the survival curves varies between recurrences. Neglecting such characteristics of non-proportionality and stratification of recurrences may yield inefficiency by encumbering borrowing strength from potentially related information sources, and consequently may jeopardize the prediction of the global survival times. Moreover, dependency might exist among the treatment strata and the stratification of recurrences [5, 9].

Consequently, more complex modeling is needy to characterize the dependence among the baseline hazard functions and treatment strata due to the temporal effects of recurrences. Frequentist inference and computing are pretty challenging and even infeasible within the most complex model setting. In Bayesian literature, there exists work that allows the baseline survival/hazard function to vary on a single level such as subgroups or the time axis ([10, 11]; among others). Nevertheless, rare work has taken bi-level varying baseline survival/hazard function into account [12], not to mention that dependence among treatment strata [5].

We propose a generalized shared frailty model (GSFM) for multiple events time data that allows the baseline hazard function to change dually along with the types of events and treatment strata, so as to strengthen the ability to borrow information from many sources. The proposed GSFM postulates multiplier frailties including both parametric and nonparametric ones, where the parametric frailty random effect accounts for the within-subject association by treating each subject as a cluster; and a nonparametric frailty effect represents dependency among treatment strata and temporal recurrences. For the GSFM, we suggest a Bayesian solution to estimate the regression coefficient vector, the variance parameter of the frailty

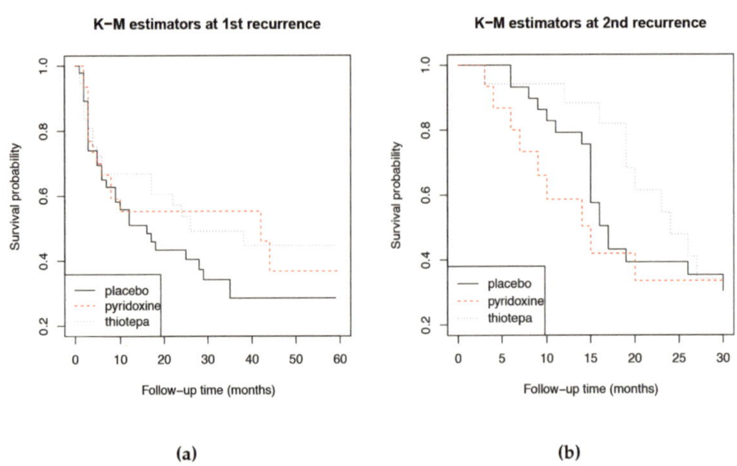

(a) (b)

Figure 1.
The Kaplan-Meier estimator of survival functions for first recurrence time (a) and second recurrence (b) in the bladder cancer data.

term, and baseline survival functions stratified by treatments and recurrences. In a Bayesian workflow, the posterior distribution is determined by the combination of observational data in the form of the likelihood function and the prior distribution represented based on the background knowledge. From a Bayesian perspective, we model the dependent nonparametric prior through transferring the data context aforementioned into the ANOVA dependent Dirichlet process (ANOVA DDP), which will be further reviewed in Section 2. The construction of the No-U-Turn sampler for Markov chain Monte Carlo (MCMC) sampling is automated by Stan [13] with its R interface [14]. The posterior inference is conducted by Stan as well.

The rest of this chapter is organized as follows. In Section 2, under typical data scenarios of dependence structure, we summarize several modified versions of the dependent Dirichlet process (DDP) initiated from MacEachern's regression spirit that nests dependent predictors into the traditional Dirichlet process (DP). In Section 3, we postulate the GSFM and transform the dependent dual-stratified multiple events to the survival-function version of the ANOVA DDP. We have a short comparison between Stan and Nimble, two contemporary Bayesian computing tools based on our user experience. In Section 4, we demonstrate the validity of the GSFM, its Bayesian inference, and analysis of the data on recurrences of bladder cancer. A brief conclusion is contained in Section 5.

2. Review of MacEachern's DDP

The DP is the most popular Bayesian nonparametric prior since the seminal work of [15]. The belief in data background that there exists some kind of dependence structure stimulates construction and selection of proper dependent prior. Some dependent DPs are constructed for unsupervised purposes such as clustering [16, 17]. The DDP prior adopted in our proposed model is supervised and predictor-dependent, originated from [18, 19], named as MacEachern's DDP in two recent review papers, which are interpretive and comprehensive [20, 21]. The key idea behind MacEachern's DDP is that the distributions of the random measures are marginally DP distributed, validated by in our Subsections 3.2 and 3.3. Therefore, we here confine how MacEachern's DDP (henceforth we use the DDP to denote the MacEachern's DDP if the context is clear) came into being expanded from the DP; and compare various modified versions of the DDP under various dependency structures. We focus on two fundamental elements of Sethuraman's construction of DP [22], the weight and the atom, following the insight of [21].

2.1 DP vs. DDP

The DP is a distribution on distributions whereas the DDP aims to construct a prior for a collection of distributions $\mathscr{F} = \{F_x | x \in \mathcal{X}\}$ indexed by covariate x. In general, there are several representations of the DP such as Polya Urn, Levy measure, and stick-breaking representations [23]. Here, we use Sethuraman's stick-breaking construction to connect the DP with the DDP. The stick-breaking construction is a kind of infinite sum representation that divides the DP into two countable series, the stick-breaking weights (SBW) and the atoms. Generally, a DP is expressed as a process with two components, the mass parameter determining the weights and the base measure to generate atoms. Through the stick-breaking construction, the DDP can be easily extended from the DP. We list their comparison in **Table 1**, where we can find that the dependency among the covariates set $\mathcal{X}$ is realized by indexing the mass parameter and base measure with the covariate $x \in \mathcal{X}$.

More specifically, the dependency can be characterized through the dependency among the weights and atoms in the DDP.

The DDP can be widely applied to scenarios of various dependence data structures. We review modification versions of the DDP from three categories depending on which part it modifies in the stick-breaking representation, weights, atoms, or both. The first is to impose the dependency on the atoms but keep common weights, leading to two typical representatives, ANOVA and Spatial [9, 24, 25]. The ANOVA type DDP encoded the covariate dependence in the form of regression for the atom processes. The Spatial DDP models for nonstrationary spatial random fields with heterogeneous variance. The second category is to modify the weights to be dependent but keep the common atoms. The early and typical work is the time series DDP [26]. They introduced a Markov Beta process on the weights to account for the temporal dependency. The third category is to impose dependency on both weights and atoms [27]. They constructed vector autoregressive and autoregressive models for atoms and weights, respectively. We summarize the aforementioned types of typical modifications in **Figure 2**.

3. Model and Bayesian inference

Consider a clinical trial with multiple event types, for example, the time of the kth recurrence of a certain disease. In the trial, n subjects are divided into G strata of treatment. Our goal is to describe the relationship between the time to the kth recurrence of a subject, and its treatment stratum, as well as its vector of covariates Z. For a certain subject, the time of recurrences, may be dependent since they occur on the same individual and thus we assume an unobservable independent shared-frailty random effect W to account for this dependence. On the other hand, we may allow the conditional hazard affiliated with the script pair kj to imply distinct survival distributions along with the temporal order of the recurrences of the disease and for specific treatment. For the ith subject in the jth treatment stratum, at the kth recurrence, given the value of frailty variable w_i and its covariate vector z_{kji}, we propose the following frailty model,

$$\lambda_{kj}\left(t|w_i, z_{kji}\right) = w_i \lambda_{0kj}(t) \exp\left(\beta^T z_{kji}\right), k = 1, \cdots, K, j = 1, \cdots, G, i = 1, \cdots, n_j. \quad (1)$$

Model (1) is called the *generalized* shared frailty model in the sense that non-proportionality among k-varying recurrences is allowed by the fact that the right-hand baseline hazard has footnotes k and j. We allow dependency among treatment strata in the model (1). Therefore, the baseline hazard function λ_{0kj} acts as if a nonparametric frailty random measure accounting for the dependency owing to the recurrences and treatment schemes.

	DP	**DDP**
Random probability measure	$F \sim \text{DP}(M, F_0)$	$\mathcal{F} = \{F_x \| x \in \mathcal{X}, M_x, F_{0x}\}$
Sethuraman's construction	$F(\cdot) = \sum_{h=1}^{\infty} p_h \delta_{\theta_h}(\cdot)$	$F_x(\cdot) = \sum_{h=1}^{\infty} p_{xh} \delta_{\theta_{xh}}(\cdot)$
	$p_h \sim \text{SBW}(1, M)$	$p_{xh} \sim \text{SBW}(1, M_x)$
	$\theta_h \sim F_0$	$\theta_{xh} \sim F_{0x}$
Convolution	$H(y) = \int k(y\|\theta) dF(\theta)$	$H_x(y) = \int k(y\|\theta) dF_x(\theta)$

Table 1.
Comparison of DP and DDP.

	DDP			
	Dependent atoms		Dependent weights and atoms	Dependent weights
	ANOVA DDP	Spatial DDP	Discrete-Time-Dependent DPM	tsDDP
DDP	$F_x(\cdot) = \sum_{h=1}^{\infty} p_h \delta_{\theta_{xh}}(\cdot)$	$G_D(\cdot) = \sum_{h=1}^{\infty} p_h \delta_{\theta_{h,D}}(\cdot)$	$F_t(\cdot) = \sum_{h=1}^{\infty} p_{th} \delta_{\theta_{th}}(\cdot)$	$F_d(\cdot) = \sum_{h=1}^{\infty} p_{dh} \theta_h(\cdot)$
Atoms	$\theta_{xh} = \alpha_h^\top d_x$ $\alpha_h \sim F_0$	$\theta_{h,D} \sim G_0$ $G_0^{(n)} \sim N_n(0, \Sigma)$	$\theta_{th} = (\mu_{th}, \Sigma_h)$ $\mu_{th} \sim N(\mu_0, V_o)$ $\mu_{t+1,h} \mid \mu_{th} \sim N(m + \Theta\mu_{th}, V)$	$\theta_h \sim F_0$
Weights	$p_h \sim \mathrm{SBW}(1, M)$	$p_h \sim \mathrm{SBW}(1, M)$	$p_{th} = (1 - \beta_{th})\Pi_{j=1}^{h=1} \beta_{tj}$ $\beta_{th} = \exp(-(2\alpha)^{-1}(\xi_h + \eta_{th})^2)$ $\xi_h, \eta_{1h} \sim N(0, 1)$ $\eta_{t+1,h} \mid \eta_{th} \sim N(\phi\eta_{th}, 1 - \phi^2)$	$p_{dh} = v_{dh}\Pi_{j=1}^{h-1}(1 - v_{dj})$ $v_{1h} \sim \mathrm{Beta}(1, M)$ $z_{dh} \mid v_{dh} \sim B(m_{dh}, v_{dh})$ $v_{d+1,h} \mid v_{dh} \sim \mathrm{Beta}(1 + z_{dh}, M - z_{dh} + m_{dh})$
Convolution	$H_x = \int N(\alpha^\top d_x, s) dF(\alpha)$ $F \sim \mathrm{DP}(M, F_0)$	$H_D = \int N(\theta_D, \sigma^2) dG(\theta_D)$ $G \sim \mathrm{DP}(M, G_0)$	$H_t = \int N(\mu, \Sigma) dF_t(\mu, \Sigma)$	No convolution
Authors	De Iorio, Müller, Rosner and MacEachern (2004, JASA)	Gelfand, Kottas and MacEachern (2005, JASA)	DeYoreo and Kottas (2018, JASA)	Nieto-Barajas, Müller, Ji, Lu and Mills (2012, Biometrics)

Figure 2.
Workflows of representative expansions of DDP.

Model (1) is an extension of the classical shared frailty model (4.1.1) on page 101 of [4] since the baseline hazard function there does not vary among the recurrences and the treatment strata. Model (1) has the analog spirit to the frailty model (1) in [10], whereas their treatment strata are independent.

3.1 Likelihood

The corresponding survival function of the model is given by: $S_{kj}(t|w_i, z_{kji}) = \{S_{0kj}(t)\}^{\exp(\beta^T z_{kji}+v_i)}$, where $S_{0kj}(t) = \exp\left(-\int_0^t \lambda_{0kj}(s)ds\right)$ denotes the baseline survival function of the kth recurrence for subjects in the jth treatment stratum, and $v_i = \log(w_i)$ denotes logarithm transformation of the frailty effect. Let f_{0kj} be the corresponding baseline density function.

Given the data sample $(Y_{kji}, \delta_{kji}, z_{ji})$, where $Y_{kji} = \min(C_{kji}, T_{kji})$, $\delta_{kji} = I(T_{kji} \le C_{kji})$, with T_{kji} being the gap time between the $(k-1)$th and kth recurrence of the ith subject in the jth stratum and C_{kij} being the corresponding censoring variable that is independent of T_{kji} given the covariate vector z_{kji}, for $k = 1, \cdots, K, j = 1, \cdots, G, i = 1, \cdots, n_j$, and $\sum_{j=1}^{G} n_j = n$. In the jth stratum, suppose that there are n_{kj} $(n_{kj} \le n_j)$ subjects suffering from the kth recurrence. Then the likelihood is written as:

$$\prod_{k=1}^{K}\prod_{j=1}^{G}\prod_{i=1}^{n_{kj}}\left[\exp(\beta^T z_{kji} + v_i) f_{0kj}(y_{kji})\left\{S_{0kj}(y_{kji})\right\}^{\left(\exp(\beta^T z_{kji}+v_i)-1\right)}\right]^{\delta_{kji}} \times \left\{S_{0kj}(y_{kji})\right\}^{(1-\delta_{kji})\exp(\beta^T z_{kji}+v_i)}.$$

3.2 Survival-function version of the ANOVA DDP

In the Bayesian workflow for the estimation, prior distributions are first determined. We here specify appropriate nonparametric priors for S_{0kj} and f_{0kj}. Since they can be easily derived from one to the other, we here only introduce the priors for S_{0kj}.

We divide S_{0kj} into K groups, and the kth group has G baseline survival functions of different treatment strata at the kth time of recurrence. That is, for a fixed k, $S_k = \{S_{0kj}, j = 1, \cdots, G\}$ is a collection of baseline survival functions with length G indexed by the categorical covariate j denoting the treatment stratum. The next procedures come from the spirit of [9]. As a general example, suppose two dugs A and B will be taken in treatment, with V and U levels of doses, respectively. In this case, $G = VU$ denotes the number of treatment strata and let the level of the jth stratum be (v,w). We write the stick-breaking form of S_{0kj} such that $S_{0kj}(t) = \sum_{h=1}^{\infty} p_h I(t > \theta_{kjh})$. We impose an ANOVA structure on θ_{kjh}:

$$\theta_{kjh} = m_{kh} + A_{kvh} + B_{kwh}, \tag{2}$$

where m_{kh} denotes the ANOVA effect shared by all the strata at the kth recurrence, and the rest terms are the ANOVA effects of the jth stratum at the kth recurrence. Let the three components be independently generated from three distributions, and marginally on j, the baseline survival function S_{0kj} follows a DP. The aforementioned procedure implies that S_k is a survival-function version of the

ANOVA DDP. If we consider a single event, that is, $K = 1$, with linear effects of a regressor vector z involved in (2) such that $\theta_{jh} = m_h + A_{vh} + B_{wh} + \beta_h^T Z$, then it reduces to the univariate survival regression model in [25].

Since any function in the stick-breaking form is discrete almost surely, we place a convolution through the Dirichlet process mixture (DPM) model [28]. Particularly, since the baseline survival functions are defined on the positive half-real line, the convolution kernel in DPM should be positive such as log-normal, Gamma, and Weibull. In this chapter, a log-normal kernel is considered. For different recurrences, we treat the relationship among $\mathcal{S}_k s$ to be independent.

3.3 One-way ANOVA DDP

Considering the data of our interest, where only one drug and one level of dose is used in each treatment stratum, we introduce the modeling of the survival-function version of one-way ANOVA DDP. In this case, the prior for the $\mathcal{S}_k$ reduces to a one-way ANOVA form since the dependency among the G treatment strata is explained by only one ANOVA effect. Furthermore, if we set $m_{kh} = 0$, $\alpha_{kh} = (\theta_{k1h}, \cdots, \theta_{kGh})^T$ reduces to a G-variate variable denoting the locations of all G baseline distributions and thus $\theta_{kjh} = \alpha_{kh}^T d_j$, where d_j is the design vector of the jth stratum to select the appropriate ANOVA effects corresponding to j.

With the above notations, we summarize the procedure to construct the survival-function version of one-way ANOVA DDP prior in model (1) as follows:

1. Stick-breaking form. For $k = 1, \cdots, K$, let $\mathcal{H}_k$ be the collection of G distribution functions s.t $\mathcal{H}_k = \{H_{kj}, j = 1, \cdots, G\}$. $H_{kj}(\cdot) = \sum_{h=1}^{\infty} p_{kh} \delta_{\theta_{kjh}}(\cdot)$.

2. Convolution step. Let $\alpha_{kh} = (\theta_{k1h}, \cdots, \theta_{kGh})^T$, and d_j be the jth design vector of length G with the jth element being 1 and others being 0. Let $H_{0k} = (H_{0k1}, \cdots, H_{0kG})$ be the collection of base measures, $S_{0kj}(t) = \int S_{\text{LN}}(t|\alpha_k^T d_j, \sigma^2) dH_k(\alpha, \sigma)$, where S_{LN} denotes the survival function of the log-normal distribution, and $H_k \sim \text{DP}(M_k, H_{0k})$.

3. Determine the mass parameter and the base measure. For simplicity, we set $M_k = 1$ for all k, which is a commonly used default value of the mass parameter [29], $H_{0k}(\theta, \sigma) = N(0, I_G) \times \text{Cauchy}(0, 5)^+$, where Cauchy$^+$ denotes the half_Cauchy distribution.

Step 1 is a standard stick-breaking representation for DP. Step 2 is kernel mixture of DP whereas the kernel is a survival function rather than a cumulative distribution function. The realization of Step 2 is quite straightforward in Stan as it provides the function lognormal_lccdf to be used as the kernel of the survival function of the log-normal family.

In Step 3, we specify the base measure as the prior for the location and shape parameters of the log-normal kernel directly rather than adding another hyper prior distribution like [25] did. The main reason is to simplify the computation in Stan. Particularly, inspired by [30, 31], we use the half-Cauchy distribution as the non-informative prior for the variance parameter instead of the inverse Gamma prior. In our practice, the choice of half-Cauchy prior significantly improves the speed of convergence and mixture performance of the MCMC chains in our real data analysis and simulation. Another interesting point we met in numerical studies is that the informativeness of the base measure for θ. Here, we do not assign the non-informative distribution but a weakly informative one is considered since we find such a weakly informative prior provides better MCMC performance than that of non-informative one with a higher effective sample size and better mixture performance. In our other research experience, the weakly informative prior for the variance parameter in the mixing component of the DPM seems to be more preferable.

3.4 Other priors and MCMC

In terms of the prior for the parametric prior w_i, we choose to log-normal prior that $v_i = \log(w_i)$ and $v_i \sim N(0, \tau^2)$, where $\tau > 0$ is an unknown parameter. We further assign a half Cauchy prior for τ s.t $\tau \sim \text{Cauchy}^+(0, 5)$ as a non-informative prior. The prior for the vector of regression coefficients is $\beta \sim N(0, 1000I)$ as a non-informative prior.

We use the truncated Dirichlet process to replace the infinite summand in the DP. The selection of the truncation point is often ad-hoc. Since in Stan the NUTS cannot sampler discrete parameters, we have to fit the truncation number and the mass parameter before the MCMC procedure. In general, the truncation number is set to be large enough s.t the truncated part is negligible. Gelman et al. [32] suggests using a truncation number L that is greater than $5M + 5$. In our computation, we set $L = 12$.

The MCMC sampling for the posterior distribution is realized in Stan. Stan and its **R** version are widely used in statistical modeling and high-performance statistical computing, especially in Bayesian. Stan realizes the MCMC sampling through the No-U-Turn sampler (NUTS). Stan automates the deriving of the fully conditional posterior distribution and NUTS is able to obtain high effective sample size [33].

3.5 Stan and NIMBLE: Programming styles

The MCMC sampling procedure is implemented in Stan and we also tried to implement the model in NIMBLE, another contemporary Bayesian computing tool in **R**. Stan and NIMBLE are two contemporary Bayesian computing tools that have drawn arising interest for Bayesian analysis but still remain under active development [34, 35]. The main advantage of Stan and NIMBLE is that they provide clear automatic posterior sampling procedures based on their specific sampling algorithms without particular justification. Therefore, users can be released from complicated probabilistic deriving and implementation. There has been a buzz group discussion about the comparison between Stan and NIMBLE in environments like [36, 37]. One comparison of their built-in samplers is demonstrated through implementing weakly informative and informative estimation within the trimmed mean regression model setting [38]. Here we contribute a naive comparison on their programming styles based on the first two authors' experience in coding this project and using Stan and NIMBLE, respectively.

A Bayesian paradigm is made up of three main steps, the prior, likelihood, and the posterior. MCMC generates samples to approximate the posterior distribution. Therefore, what one needs to set in a Bayesian computing tool is the prior and likelihood, let alone Stan or NIMBLE. Nevertheless, Stan and NIMBLE take different programming styles in writing likelihood. In Stan, the default way to present the log-likelihood is the syntax target and users can add log contribution to it freely, which is similar to the natural language and straightforward to users whatever level of mathematical background. In NIMBLE, the default way is to transfer the likelihood into some standard distributions given by NIMBLE, which may not be friendly for users who have a relatively less mathematical background.

We take fitting the finite mixture of the Gaussian model as an example. For a fixed positive integer L, the distribution of Y is given by $F_Y(s) = \sum_{l=1}^{L} p_l N(s|\mu_l, \sigma_l^2)$ and the log-likelihood is $\log L(p, \mu, \sigma|Y) = \sum_{i=1}^{n} \sum_{l=1}^{L} \{ \log(p_l) + \log \phi(y_i|\mu_1, \sigma_l) \}$, where ϕ denotes the density function of normal distribution. The code for Stan and NIMBLE to implement this model is listed in Listing 1.1 and 1.2, respectively. In Listing 1.1 we clearly find that the contribution to the syntax target is just the sum of $\log(p_l)$ and the logarithm of the density of normal distribution denoted by normal_lpdf. The rest is to assign a Dirichlet prior to the weights p_l and other parameters. However, in the NIMBLE code shown in Listing 1.2, we have to transfer the likelihood into some sampling procedures by IMAGING that there are L clusters of random numbers, the random numbers are i.i.d Gaussian within each cluster, and the probability a random number is drawn from the lth cluster is p_l. Thereafter, the Dirichlet prior is assigned to p_ls. Such imagine matches the Bayesian philosophy but when the likelihood function becomes to be quite complicated, to understand this sampling procedure may not be easy anymore, especially for practitioners not coming from a mathematics or statistics background.

Listing 1.1: *Stan code for modeling mixture of Gaussian distribution*

```
data {
int<lower=1> N;
vector[N] y;
int<lower=1> L;
}
parameters {
simplex[L] p;
vector[L] mu;
vector<lower=0>[L] sigma;
}
model {
p ~ dirichlet(rep_vector(1, L));
mu ~ normal(0, 100);
sigma ~ cauchy(0, 2.5);
for(i in 1:N) {
vector[L] lp_i;
for(l in 1:L) {
lp_i[l] = log(p[l]) + normal_lpdf(y[i]|mu[l], sigma[l]);
}
target += log_sum_exp(lp_i);
}
}
```

Listing 1.2: *NIMBLE code for modeling mixture of Gaussian distribution*

```
NimbleCode <- nimbleCode ({
   for (i in 1:N) {
      y[i] ~ dnorm(mu_y[z[i]], tau = tau_y[z[i]])
      z[i] ~ dcat(p[1,L])
   }
   for (j in 1:L) {
```

```
        mu_y[j] ~ dnorm(0, 0.01)
        tau_y[j] ~ dgamma(0.01, 0.01)
    }
    p[1:L] ~ ddirch(alpha0[1,L])
})
NimbleData <- list(y = y)
NimbleConsts <- list(L = L, N = length(NimbleData$y), alpha0 = rep(1, L))
NimbleInits <- list(mu_y = rnorm(NimbleConsts$L), tau_y = rgamma
        (NimbleConsts$L),p = rep(1/NimbleConsts$L, NimbleConsts$L))
```

4. Application: Bladder cancer recurrences

We apply the GSFM to analyze the bladder cancer recurrences data set contained in R package survival. Totally 118 subjects in the clinical trial are divided into three treatment strata including placebo, pyridoxine (vitamin B6), and thiotepa. Each subject may experience k (from 1 to 9) times of recurrences and may die from or not from the recurrence of bladder cancer. We do not discriminate the death from cancer and the recurrence, and the death from other causes is treated as censoring status. Our interest is the gap time between the $(k-1)$th and the kth recurrences. Besides the treatment schemes, two clinical covariates are considered: the number of tumors at the beginning (x_1) and the size of the largest tumor (x_2) within a subject. The values of these two covariates are evaluated at the beginning of each recurrence interval. This data set was once analyzed for the time between the first to the second recurrence as a univariate time-to-event outcome [39]. In this chapter, we consider both the first and the second recurrences and thus $K = 2$ here. The two covariates are scaled by divided by 100. To simplify the computation, the follow-up time is transferred from months to years to get lower scalars.

4.1 Model-checking for baseline survival functions

Before further inference, we need to check whether the proposed model is appropriate. As an alternative, a shared frailty model is fit by R package spBayeSurv. In the shared frailty model, the treatment strata are considered as indicator covariates in the parametric term. We run four independent MCMC chains for 5000 times with the first 2000 times burn-in and aggregate the rest chains together as the posterior samples under the GSFM. All chains are well mixed and convergent under the GSFM. For the shared frailty model, we run the MCMC 16,000 times with the first 6000 times burn-in through R function survregbayes using the "IID" Gaussian frailty under "PH" model name. Other settings are default.

The plots of the estimated baseline survival functions under different models stratified by treatment strata can be viewed in **Figure 3**. From that, we find the baseline survival functions estimated under the GSFM show similar trends as that of the K-M estimator in each recurrence and reflect the crossing survival curves at the first recurrence like the K-M estimator. However, the curves estimated by the shared frailty model are not crossed and cannot change along with recurrences. Therefore, the proposed GSFM is appropriate for the data.

4.2 Parametric estimation I: Real data

We use the mean of posterior samples (median for τ) as the estimator of parameter and we list the estimation of the vector of regression coefficients β and standard deviation parameter τ in **Table 2**.

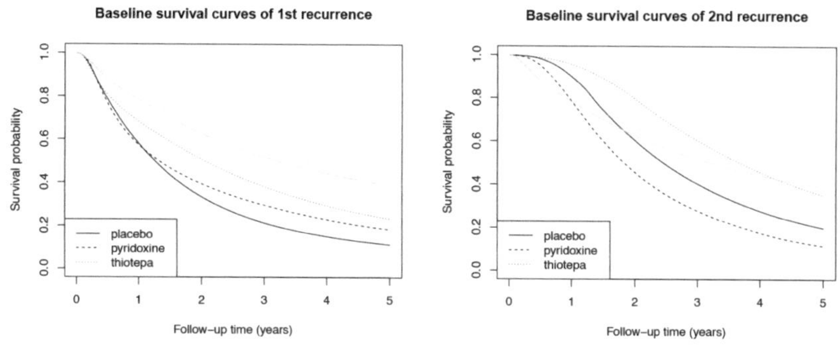

(a) (b)

Figure 3.
The estimated baseline survival curves for the first (a) and second (b) recurrence; the black curves are estimated under the proposed generalized shared frailty model, and the pink curves are estimated under the traditional shared frailty model; the real lines, placebo; the dash lines, pyridoxine; the dotted lines, thiotepa.

	Est	SD	ESS	PACE
No. tumors	13.849	11.051	1495	0.145
Tumor size	−14.196	12.341	1114	0.194
τ	1.793	0.383	456	0.474

Est, point estimation; SD, posterior standard deviation; ESS, effective sample size; PACE, the MCMC Pace.

Table 2.
The parametric estimation and the MCMC performance for the bladder cancer recurrences data.

From **Table 2** we find that as the number of tumors at the start point increases, the hazard for recurrences increases as well whereas the larger size of the largest tumor will decrease the hazard. This conclusion is similar to that in [39] who analyzed the first recurrence as univariate time-to-event data by a transformation model.

Besides the parametric estimation result, we also report two metrics about the MCMC performance here. The first one is the effective sample size (ESS), an approximation to the number of "independent" draws in MCMC sampling. It shows that the ESS of all parameters is greater than 400, which is considered to be adequate by [40]. The ESS of τ is significantly lower than that of β, a possible reason is that the frailty random effect might be time-dependent $w_i(t)$ rather than a time-fixed effect. Another metric of interest is the average time needed to generate each effective sample, called MCMC Pace. Stan development team emphasized the importance of MCMC Pace, and the definition is given by the team of NIMBLE in [41] as the time consuming of generating one effective sample. The MCMC Pace to generate τ is much higher than that of β, and we conjecture the possible reason is that the posterior distribution has a long upper tail leading to outliers in posterior samples, which slows down the speed to generate effective samples.

4.3 Parametric estimation II: Simulation

Another simulation study is considered to evaluate the performance of parametric estimation of the MCMC procedure. Our simulation aims to simulate the occurrences of multiple events on the same individual. We take $K = 2$ and $G = 3$ denote the number of types of events and the number of treatment strata,

Parameter	BIAS	RMSE	ESD	SDE	CP
$\beta_1 = 1$	−0.062	0.042	0.222	0.196	96.7
$\beta_2 = 1$	−0.025	0.023	0.148	0.152	92.7
$\tau = 1$	−0.078	0.056	0.213	0.224	96.7

BIAS, averaged bias among the 150 simulations; RMSE, the root of mean square error of the estimation; ESD, averaged posterior estimated standard deviation; SDE, the standard deviation of point estimate; CP, the coverage probability of 95% credible interval.

Table 3.
Simulation results for the parametric terms.

respectively. The simulation includes two independent covariates, $x_i \sim \text{Bin}(1, 0.5)$ and $x_2 \sim N(0, 1)$ to incorporate indicator variable and continuous variable as well. For $k = 1, 2, j = 1, 2, 3$, the baseline survival functions S_{0kj} are set as:

- $S_{011} = 1 - 0.5(LN(-0.25, 1) + LN(0.25, 1))$;
- $S_{012} = 1 - 0.5(LN(-0.5, 1) + LN(0.65, 1))$;
- $S_{013} = 1 - 0.5(LN(-0.65, 1) + LN(1.25, 1))$;
- $S_{021} = 1 - LN(0, 1)$; $S_{022} = 1 - LN(-0.5, 1)$; $S_{023} = 1 - LN(0.5, 1)$

When $k = 1$, the three baseline survival functions are crossed whereas when $k = 2$, the three curves are not. The vector of regression coefficients is $\beta = (1, 1)^T$ and the log frailty random effect $v_i \sim N(0, 1)$ independently. The survival time is generated following model (1). The censoring variable of each event is generated from Unif(4,6) independently, leading to a censoring rate of about 28%. We set the number of subjects to be 90 and they are equally divided into three treatment strata. We repeat the simulation for 150 times.

Table 3 summarizes the results for regression parameters β and the standard deviation of frailty effect τ, including the averaged bias (BIAS), the root of mean square error (RMSE), posterior estimated standard deviation (ESD) of each point estimate (posterior mean for β and median for τ), the standard deviation (across 150 replicated simulations) of the point estimate (SDE), and the coverage probability (CP) of the 95% credible interval (given by Wald-type credible interval). The results show that the point estimates of β and τ have quite little bias with low RMSE, ESD values are close to the corresponding SDEs, and the CP values are close to the nominal 95%.

5. Discussion

In this chapter, we show the power of Bayesian computing illustrated by successfully applying the ANOVA DDP model as the nonparametric prior for a relatively complicated shared frailty model. Our survival-function version of the ANOVA DDP, modified based on the ANOVA DDP directly in Subsection 3.3, is constructed for the shared frailty model, but can reduce to modeling the univariate dependent survival functions by involving the continuous covariates into the predictor space of the ANOVA DDP. Hence, our work is an extension of [25] to some extent. However, the proposed GSFM is different from the Linear DDP model for a single group which is a generalization of the accelerated failure time model [42, 43].

Furthermore, although we point out that there exists potentially dual dependence for dual stratification of treatment strata and recurrences, we just simply allow dependence in treatment strata and assume that the recurrences are independent in our methodology demonstration. The dependence across recurrences per subject is dealt with only by the parametric frailty random effect in the proposed shared frailty model. It is more reasonable to be incorporated into the baseline survival functions so that the interaction effects between recurrence and treatment may be accounted for. Under the one-level stratification, Hanson et al. [5] modeled such serial correlation among baseline hazard functions by constructing the so-called dependent tail free process as the prior. It is non-trivial to accommodate dual temporal and stratified dependency as a future research plan.

Acknowledgements

Chong Zhong's research was partially supported by GRF1531519, RGC, HKSAR. Zhihua Ma's research was partially supported by Shenzhen Institutions Stability Support Program 20200812101943002, China. Junshan Shen's research is partially supported by the Beijing Natural Science Foundation 1192006, China. Catherine Liu's research was partially supported by HKPOLYU grant YBTR, and GRF1531519, RGC, HKSAR.

Thanks

The first author Chong Zhong owes deep thanks to his parents. The authors thank Miss. Lulu Zhang for her efficient technical supports in text and figures. The authors thank the service manager Ms. Romina Rovan for her courtesy and professional service. The authors thank the invitation from the editor.

Author details

Chong Zhong[1†], Zhihua Ma[2†], Junshan Shen[3†] and Catherine Liu[1*†]

1 The Hong Kong Polytechnic University, Hong Kong, China

2 Shenzhen University, Guangdong, China

3 Capital University of Economics and Business, Beijing, China

*Address all correspondence to: catherine.chunling.liu@polyu.edu.hk

† These authors contributed equally.

References

[1] Vaupel JW, Manton KG, Stallard E. The impact of heterogeneity in individual frailty on the dynamics of mortality. Demography. 1979;**16**(3): 439-454

[2] Duchateau L, Janssen P. The Frailty Model. New York: Springer Science & Business Media; 2007

[3] Balan TA, Putter H. A tutorial on frailty models. Statistical Methods in Medical Research. 2020;**29**(11): 3424-3454

[4] Ibrahim JG, Chen M-H, Sinha D. Bayesian Survival Analysis. New York: Springer Science & Business Media; 2001

[5] Hanson TE, Jara A, Zhao L, et al. A bayesian semiparametric temporally-stratified proportional hazards model with spatial frailties. Bayesian Analysis. 2012;**7**(1):147-188

[6] de Castro M, Chen M-H, Zhang Y. Bayesian path specific frailty models for multi-state survival data with applications. Biometrics. 2015;**71**(3): 760-771

[7] Therneau TM. A Package for Survival Analysis in R. R Package Version 3.2-11. 2021

[8] Zeng D, Lin DY. Maximum likelihood estimation in semiparametric regression models with censored data. Journal of the Royal Statistical Society, Series B: Statistical Methodology. 2007; **69**(4):507-564

[9] De Iorio M, Müller P, Rosner GL, MacEachern SN. An anova model for dependent random measures. Journal of the American Statistical Association. 2004;**99**(465):205-215

[10] de Castro M, Chen M-H, Ibrahim JG, Klein JP. Bayesian transformation models for multivariate survival data. Scandinavian Journal of Statistics. 2014; **41**(1):187-199

[11] Paulon G, De Iorio M, Guglielmi A, Ieva F. Joint modeling of recurrent events and survival: A bayesian non-parametric approach. Biostatistics. 2020;**21**(1):1-14

[12] Conlon ASC, Taylor JMG, Sargent DJ. Multi-state models for colon cancer recurrence and death with a cured fraction. Statistics in Medicine. 2014; **33**(10):1750-1766

[13] Stan Development Team. The Stan Core Library. Version 2.27. 2018

[14] Stan Development Team. RStan: The R Interface to Stan. R Package Version 2.21.2. 2020

[15] Ferguson TS. Prior distributions on spaces of probability measures. The Annals of Statistics. 1974;**2**(4):615-629

[16] Teh YW, Jordan MI, Beal MJ, Blei DM. Hierarchical dirichlet processes. Journal of the American Statistical Association. 2006;**101**(476):1566-1581

[17] Rodriguez A, Dunson DB, Gelfand AE. The nested dirichlet process. Journal of the American Statistical Association. 2008;**103**(483):1131-1154

[18] MacEachern SN. Dependent nonparametric processes. In: ASA Proceedings of the Section on Bayesian Statistical Science. Alexandria, VA: American Statistical Association; 1999

[19] MacEachern SN. Dependent Dirichlet Processes. Technical Report. Department of Statistics, The Ohio State University; 2000

[20] MacEachern SN. Nonparametric bayesian methods: A gentle introduction

and overview. Communications for Statistical Applications and Methods. 2016;**23**(6):445-466

[21] Quintana FA, Mueller P, Jara A, MacEachern SN. The dependent dirichlet process and related models. arXiv preprint arXiv:2007.06129. 2020

[22] Sethuraman J. A constructive definition of Dirichlet priors. Statistica Sinica. JSTOR. 1994;**4**:639-650

[23] Phadia EG. Prior Processes and their Applications. New York: Springer; 2015

[24] Gelfand AE, Kottas A, MacEachern SN. Bayesian nonparametric spatial modeling with dirichlet process mixing. Journal of the American Statistical Association. 2005;**100**(471):1021-1035

[25] De Iorio M, Johnson WO, Müller P, Rosner GL. Bayesian nonparametric nonproportional hazards survival modeling. Biometrics. 2009;**65**(3): 762-771

[26] Nieto-Barajas LE, Müller P, Ji Y, Lu Y, Mills GB. A time-series ddp for functional proteomics profiles. Biometrics. 2012;**68**(3):859-868

[27] DeYoreo M, Kottas A. Modeling for dynamic ordinal regression relationships: An application to estimating maturity of rockfish in california. Journal of the American Statistical Association. 2018;**113**(521): 68-80

[28] Lo AY. On a class of bayesian nonparametric estimates: I. Density estimates. The Annals of Statistics. 1984; **12**:351-357

[29] Gelman A, Carlin JB, Stern HS, Dunson DB, Vehtari A, Rubin DB. Bayesian Data Analysis. Florida: CRC Press; 2013

[30] Gelman A. Prior distributions for variance parameters in hierarchical models (comment on article by browne and draper). Bayesian Analysis. 2006; **1**(3):515-534

[31] Gelman A, Jakulin A, Pittau MG, Su Y-S. A weakly informative default prior distribution for logistic and other regression models. The Annals of Applied Statistics. 2008;**2**(4):1360-1383

[32] Ohlssen DI, Sharples LD, Spiegelhalter DJ. Flexible random-effects models using bayesian semi-parametric models: Applications to institutional comparisons. Statistics in Medicine. 2007;**26**(9):2088-2112

[33] Hoffman MD, Gelman A, et al. The no-u-turn sampler: Adaptively setting path lengths in hamiltonian monte carlo. Journal of Machine Learning Research. 2014;**15**(1):1593-1623

[34] Kerioui M, Mercier F, Bertrand J, Tardivon C, Bruno R, Guedj J, et al. Bayesian inference using hamiltonian monte-carlo algorithm for nonlinear joint modeling in the context of cancer immunotherapy. Statistics in Medicine. 2020;**39**(30):4853-4868

[35] Ma Z, Hu G, Chen M-H. Bayesian hierarchical spatial regression models for spatial data in the presence of missing covariates with applications. Applied Stochastic Models in Business and Industry. 2021;**37**(2):342-359

[36] Stan forums. Available from: https://discourse.mc-stan.org

[37] Nimble groups. Available from: https://r-nimble.org/more/issues-and-groups

[38] Zhang L. A bayesian comparison in stan and nimble by trimmed mean regression [M.Phil's thesis]. Hong Kong, China: The Hong Kong Polytechnic University; 2021

[39] Zeng D, Lin DY. Efficient estimation of semiparametric

transformation models for counting processes. Biometrika. 2006;**93**(3): 627-640

[40] Vehtari A, Gelman A, Simpson D, Carpenter B, Bürkner P-C. Rank-normalization, folding, and localization: An improved r for assessing convergence of MCMC. Bayesian Analysis. 2021;**1**(1):1-28

[41] de Valpine P, Paciorek C, Turek D, Michaud N, Anderson-Bergman C, Obermeyer F, Cortes CW, RodrÃguez A, Lang DT, Paganin S. NIMBLE User Manual. R Package Manual Version 0.11.1. 2021

[42] Hanson TE, Jara A. Surviving fully bayesian nonparametric regression models. Bayesian Theory and Applications. 2013:593-615

[43] Riva-Palacio A, Leisen F, Griffin J. Survival regression models with dependent bayesian nonparametric priors. Journal of the American Statistical Association. 2021:1-10

Chapter 6

Modeling Heterogeneity Using Lindley Distribution

Arvind Pandey and Lalpawimawha

Abstract

Frailty models are intended for use in survival analysis to explain unobserved heterogeneity in an individual caused by various hereditary variables or environmental influences. A shared frailty model was utilized to examine the data. It is based on the idea that frailty affects the hazard rate in a multiplicative manner. In this manuscript, we introduce a new frailty model called the Lindley shared frailty model with exponential power and generalized Rayleigh as baseline distributions. The The Bayesian method of the Monte Carlo method of the Markov chain is used to estimate the parameters used in the model; simulation studies are also carried out to compare the actual and calculated values of the parameters; the proposed model is compared with the Bayesian comparison method Compare and propose the best model of infectious disease data.

Keywords: Bayesian technique, exponential power distribution, generalized Rayleigh distribution, Lindley frailty, MCMC

1. Introduction

The term frailty was coined by Vaupel et al. [1]. The frailty model is typically represented as an unobservable random variable that multiplies the risk function, with the frailty random variable supposed to be one of the parameter distributions, such as gamma, log-normal, positive stable, inverse Gaussian, power variance function, and so on. Let Y be a continuous random variable of lifetime of an individual and the frailty random variable (RV) be V. The conditional hazard function (CHF) for a given frailty variable $V = v$ at time $y > 0$ is

$$m(y|v) = vh_0(y)e^{X'\beta}, \tag{1}$$

where $m_0(y)$ is a baseline hazard function (BHF) at time $y > 0$, X is a covariate and β is a regression coefficient, these are in vector form. The CHF for given frailty at time $y > 0$ is

$$S(y|v) = e^{-\int_0^y m(x|v)dx} = e^{-vM_0(y)e^{X'\beta}}, \tag{2}$$

where $M_0(y)$ is cumulative baseline hazard function (CBHF) at time $y > 0$. Integrating over the range of frailty variable V having density $f(v)$, we get marginal survival function as

$$S(y) = \int_0^\infty S(y|v)f(v)dv, \tag{3}$$

$$= L_v\left(M_0(y)e^{X'\beta}\right), \tag{4}$$

where $L_V(.)$ is a Laplace transformation of the distribution of V. Once we have survival function at time $y > 0$ of lifetime random variable of an individual one can obtain probability structure and can base their inference on it.

Frailty models have gained more attention in the recent medical research due to the uniqueness property of the frailty parameter. Generally, gamma distribution, log-normal distribution and inverse Gaussian distributions are the most commonly used frailty distributions [2, 3]. Hanagal and Dabade [4] introduced new Compound negative binomial shared frailty models for bivariate survival data using Weibull and generalized exponential as baseline distributions. Pandey et al. [5] compared gamma, inverse Gaussian and positive frailty models with generalized Pareto as baseline distribution. Pandey et al. [6] also compared gamma and inverse Gaussian frailty distributions under additive property.

To extract the features of the Lindley shared frailty model, we used Lindley as a frailty distribution with right censored data under generalized Rayleigh and exponential power as baseline distributions. The survival periods are dependent in this case because the frailty variable follows the Lindley distribution. The predicted value of the frailty distribution variance influences the population's degree of heterogeneity. The higher the variance of the frailty distribution, the more heterogeneity there is in the population under consideration. The frailty distribution becomes degraded when zero variance is observed. As a baseline distribution, the exponential power distribution is used. Because it exhibits a rising hazard rate, which is typical in real-life distributions, the exponential power distribution is chosen as the baseline distribution. The Lindley distribution with one parameter was first proposed by Lindley [7] for analyzing failure times data. It belongs to an exponential family, but it is used as an alternative to the exponential distribution. Lindley distribution is alluring due to the ability of modeling failure time data with increasing, decreasing, unimodal and bathtub shaped hazard rates. Ghitany et al. [8, 9] discussed different properties of Lindley distribution and also showed that Lindley distribution is better than the exponential distribution for modeling failure time data when considering hazard rate is unimodal or bathtub shaped. It is also shown that Lindley distribution is more flexible than exponential distribution in modeling lifetime data. Many authors have discussed and introduced different generalization of Lindley distribution. Bakouch et al. [10] introduced extended Lindley distribution. Ghitany et al. [11] proposed the power Lindley distribution and Shanker et al. [12] proposed two parameter Lindley distribution, which could also be reduced to one parameter case. The mean of a two parameter Lindley distribution is always greater than the mode indicating that the distribution is positively skewed.

The classic approach and the Bayesian approach are two widely utilized techniques in general. We can employ prior distributions here, therefore we'll estimate the model parameters using the Bayesian Markov Chain Monte Carlo (MCMC) approach. Furthermore, because characteristics with diverse posterior distributions may be easily generated, the results and model selection criteria can be clearly interpreted. Run after thinning mean and autocorrelation plots, follow-up plots, past plot couplings, sample autocorrelation plots dictate chain behavior, burn duration, autocorrelation delay, and how observations are made It's utilized for cognitive confirmation on its own. We also give simulation experiments to back up the model's performance. All of the model's estimation processes are detailed, as well as infection statistics relating to kidney infections.

In Sections 2 and 3, the introduction of the Lindley shared frailty model and baseline distributions are given, followed by proposed models and estimation strategies in Sections 4 and 5. In Sections 6 and 7, application of the proposed model and discussion are given.

2. Lindley shared frailty model

Let a continuous random variable V follows two parameter Lindley distribution (TPLDP) with parameters α and λ then density function of V is

$$f(v) = \begin{cases} \dfrac{\alpha^2}{\alpha\lambda + 1}(\lambda + v)e^{-\lambda v} & ;\ v > 0, \alpha > 0, \lambda\alpha > -1 \\ 0 & ;\ otherwise, \end{cases} \tag{5}$$

and the Laplace transform is

$$L_V(s) = \frac{\alpha^2(1 + (s + \alpha)\lambda)}{(s + \alpha)^2(1 + \lambda\alpha)}, s + \alpha > 0. \tag{6}$$

The mean and variance of frailty variable are $\mathbf{E}(Z) = \frac{\alpha\lambda+2}{\alpha(\alpha\lambda+1)}$ and $\mathbf{V}(V) = \frac{2+4\alpha\lambda+\alpha^2\lambda^2}{\alpha^2(\alpha\lambda+1)^2}$. For identifiability, we assume V has expected value equal to one i.e. $E(V) = 1$, which imply that $\alpha = \xi$ and $\lambda=\frac{2-\xi}{\xi(\xi-1)}$. Under this restriction the density function and the Laplace transformation of Lindley distribution reduces to

$$f(v) = \begin{cases} \dfrac{e^{-\xi v}\xi(\xi^2 v + \xi - \xi v - 2)}{\xi - 2} & ;\ v > 0, \xi > 0 \\ 0 & ;\ otherwise, \end{cases} \tag{7}$$

and the Laplace transform is

$$L_V(s) = \frac{\xi(\xi + s(2 - \xi))}{(s + \xi)^2}. \tag{8}$$

with variance of V is $\frac{4\xi-\xi^2-2}{\xi^2}$. The frailty variable V is degenerate at $V = 1$. Replacing Laplace transform in Eq. (4), we get the unconditional bivariate survival function for the jth individual as

$$S(y_{1k}, y_{2k}) = \frac{\xi(\xi + \eta_k(M_{01}(y_{1k}) + M_{02}(y_{2k}))(2 - \xi))}{(\eta_k(M_{01}(y_{1k}) + M_{02}(y_{2k})) + \xi)^2} \tag{9}$$

where $M_{01}(y_{1k})$ and $M_{02}(y_{2k})$ are the cumulative baseline hazard functions of the lifetime Y_{1k} and Y_{2k} respectively.

And for without frailty, the model becomes

$$S(y_{1k}, y_{2k}) = e^{-\eta_k(M_{01}(y_{1k}) + M_{02}(y_{2k}))}. \tag{10}$$

3. Baseline distributions

As a starting point, we'll look at the generalized Rayleigh distribution. Surles and Padgett [13] proposed the two-parameter Burr type X distribution, dubbed the generalized Rayleigh distribution, and demonstrated that the two-parameter generalized Rayleigh distribution may be utilized to describe strength and general lifetime data rather efficiently. The two-parameter generalized Rayleigh distribution can be utilized well in survival analysis to describe strength data as well as general lifetime data. If a continuous random variable Y has a two-parameter generalized Rayleigh distribution, the survival function, hazard function, and cumulative hazard function are as follows:

$$S(y) = 1 - \left(1 - e^{-(\lambda y)^2}\right)^{\alpha}; y > 0, \lambda > 0, \alpha > 0 \tag{11}$$

$$m(y) = \frac{2\alpha\lambda^2 y e^{-(\lambda y)^2}\left(1 - e^{-(\lambda y)^2}\right)^{\alpha-1}}{1 - \left(1 - e^{-(\lambda y)^2}\right)^{\alpha}}; y > 0, \lambda > 0, \alpha > 0 \tag{12}$$

$$M(y) = -log\left[1 - \left(1 - e^{-(\lambda y)^2}\right)^{\alpha}\right]; y > 0, \lambda > 0, \alpha > 0 \tag{13}$$

where α and λ stands for shape and scale parameters respectively of the distribution. It has also some attractive properties increasing hazard and bathtub type depends on the parameter value.

The second baseline distribution considered here is exponential power distribution. A continuous random variable Y is said to follow the exponential power distribution if its survival function, hazard function and cumulative hazard function are, respectively,

$$S(y) = e^{1-e^{\lambda y^{\alpha}}}; y > 0, \lambda > 0, \alpha > 0 \tag{14}$$

$$m(y) = \alpha\lambda y^{\alpha-1} e^{\lambda y^{\alpha}}; y > 0, \lambda > 0, \alpha > 0 \tag{15}$$

$$M(y) = e^{\lambda y^{\alpha}} - 1 \tag{16}$$

where λ and α are the shape and scale parameters of the exponential power distribution. The hazard function and cumulative hazard function are respectively,

$$m(y) = \alpha\lambda y^{\alpha-1} e^{\lambda y^{\alpha}}; y > 0, \lambda > 0, \alpha > 0 \tag{17}$$

$$M(y) = e^{\lambda y^{\alpha}} - 1 \tag{18}$$

The hazard function is decreasing function at time y when $\alpha < 1$ for smaller values of λ but as λ increases hazard function takes U shape curve and further increment in λ gives increasing nature to hazard function.

4. Proposed models

The unconditional survival function is obtained by replacing the cumulative hazard functions of generalized Rayleigh distribution and exponential power distribution in Eqs. (9) and (10). Then,

$$S(y_{1k}, y_{2k}) = e^{-\left(\left(-\log\left[1-\left(1-e^{-(\lambda_1 y_{1k})^2}\right)^{\alpha_1}\right]\right)+\left(-\log\left[1-\left(1-e^{-(\lambda_2 y_{2k})^2}\right)^{\alpha_2}\right]\right)\right)\eta_k} \left[1+\xi\left(\left(-\log\left[1-\left(1-e^{-(\lambda_1 y_{1k})^2}\right)^{\alpha_1}\right]\right)+\left(-\log\left[1-\left(1-e^{-(\lambda_2 y_{2k})^2}\right)^{\alpha_2}\right]\right)\right)\right]^{-1/\xi} \tag{19}$$

$$S(y_{1k}, y_{2k}) = e^{-\left(\frac{\alpha_1}{\lambda_1}\left(e^{\lambda_1 y_{1k}}-1\right)+\frac{\alpha_2}{\lambda_2}\left(e^{\lambda_2 y_{2k}}-1\right)\right)\eta_k} \left[1+\xi\left(\frac{\alpha_1}{\lambda_1}\left(e^{\lambda_1 y_{1k}}-1\right)+\frac{\alpha_2}{\lambda_2}\left(e^{\lambda_2 y_{2k}}-1\right)\right)\right]^{-1/\xi} \tag{20}$$

$$S(y_{1k}, y_{2k}) = e^{-\left(\left(-\log\left[1-\left(1-e^{-(\lambda_1 y_{1k})^2}\right)^{\alpha_1}\right]\right)+\left(-\log\left[1-\left(1-e^{-(\lambda_2 y_{2k})^2}\right)^{\alpha_2}\right]\right)\right)\eta_k} \tag{21}$$

$$S(y_{1k}, y_{2k}) = e^{-\left(\frac{\alpha_1}{\lambda_1}\left(e^{\lambda_1 y_{1k}}-1\right)+\frac{\alpha_2}{\lambda_2}\left(e^{\lambda_2 y_{2k}}-1\right)\right)\eta_k} \tag{22}$$

The Eqs. (19) and (20) are Lindley shared frailty model with generalized Rayleigh and exponential power as baseline distributions, called as Model-I and Model-II and Eqs. (21) and (22) are without frailty models under the same baseline distributions, called as Model-III and Model-IV.

5. Estimation strategies

By assuming independence between censoring scheme and individual lifetimes, the likelihood function associated with failure times for the kth people (k = 1,2,3, n) and censoring times is given by

$$l(\underline{\Psi}, \beta, \xi) = \prod_{k=1}^{n_1} f_1(y_{1k}, y_{2k}) \prod_{k=1}^{n_2} f_2(y_{1k}, d_{2k}) \prod_{k=1}^{n_3} f_3(d_{1k}, y_{2k}) \prod_{k=1}^{n_4} f_4(d_{1k}, d_{2k}) \tag{23}$$

where $\underline{\Psi}$, β and ξ are vectors of baseline parameters, regression coefficients and frailty distribution parameter. The likelihood function for without frailty is given as

$$l(\underline{\Psi}, \beta) = \prod_{k=1}^{n_1} f_1(y_{1k}, y_{2k}) \prod_{k=1}^{n_2} f_2(y_{1k}, d_{2k}) \prod_{k=1}^{n_3} f_3(d_{1k}, y_{2k}) \prod_{k=1}^{n_4} f_4(d_{1k}, d_{2k}) \tag{24}$$

and n_1, n_2, n_3 and n_4 are the number of observations, which are observed to lie in the intervals $y_{1k} < d_{1k}, y_{2k} < d_{2k}$; $y_{1k} < d_{1k}, y_{2k} > d_{2k}$; $y_{1k} > d_{1k}, y_{2k} < d_{2k}$ and $y_{1k} > d_{1k}, y_{2k} > d_{2k}$ respectively and the contribution of the kth individual in the likelihood function as

$$\begin{aligned} f_1(y_{1k}, y_{2k}) &= \frac{\partial^2 S(y_{1k}, y_{2k})}{\partial y_{1k} \partial y_{2k}} \\ f_2(y_{1k}, d_{2k}) &= -\frac{\partial S(y_{1k}, d_{2k})}{\partial y_{1k}} \\ f_3(d_{1k}, y_{2k}) &= -\frac{\partial S(d_{1k}, y_{2k})}{\partial y_{2k}} \\ f_4(d_{1k}, d_{2k}) &= S(d_{1k}, d_{2k}) \end{aligned} \tag{25}$$

Putting Eq. (24) in Eqs. (23) and (24), we get the likelihood functions for the Lindley shared frailty models under generalized Rayleigh and exponential power baseline distributions and likelihood function for without frailty models under the same baseline distributions.

The joint posterior density of the parameters given failure times is given as

$$\pi(\alpha_1, \lambda_1, \alpha_2, \lambda_2, \xi, \beta) \propto L(\alpha_1, \lambda_1, \alpha_2, \lambda_2, \xi, \beta)$$
$$\times g_1(\alpha_1) g_2(\lambda_1) g_3(\alpha_2) g_4(\lambda_2) g_5(\xi) \prod_{i=1}^{5} p_i(\beta_i)$$

where $g_i(.)$ $(i = 1, 2, \cdots, 5)$ represent the prior density function of baseline parameters and frailty variance, which are suppose to have known hyper parameters; $p_i(.)$ represents prior density function for the regression coefficient β_i; β_i represents regression coefficients of vector form except β_i, $i = 1, 2, \ldots, a$ and likelihood function $I(.)$ is also presented by Eqs. (23) and (24). It is assumed that all of the parameters are distributed independently in this case.

The expression of the likelihood function in Eqs. (23) and (24) are not easy to solve by using the Newton–Raphson method. MLEs fail to converge as it involved a large number of parameters. As a result, the Bayesian approach was used to estimate the parameters involved in the models, which is free of such issues.

Prior distributions are utilized as follows: for a frailty parameter with a small value of Ψ, a gamma distribution with mean 1 and big variance $\Gamma(\Psi, \Psi)$ is used as a prior distribution. As a prior for the regression coefficient, say φ^2, a normal distribution with mean zero and huge variance is utilized. Because we do not know anything about the baseline parameters, we use the same type of prior distributions used by Ibrahim et al. [14] and Sahu et al. [15], as well as a non-informative prior. As non-informative prior distributions, $\Gamma(a_1, b_1)$ and $U(a_2, b_2)$ are utilized. All the hyper-parameters $\Psi, \varphi, a_1, a_2, b_1$ and b_2 are supposed to be known in advanced. Here $\Gamma(a_1, b_1)$ stands for gamma distribution having shape parameter a_1 and scale parameter b_1 and $U(a_2, b_2)$ stands for the uniform distribution over the interval a_2 to b_2. We provide the hyper-parameters as $\Psi = 0.0001, \varphi^2 = 1000, a_1 = 1, b_1 = 0.0001, a_2 = 0$, and $b_2 = 100$.

The Metropolis Hasting Algorithm and Gibbs Sampler were used to estimate the parameters in the models fitted with the preceding prior density function and likelihood Eqs. (23) and (24), Metropolis Hasting Algorithm and Gibbs Sampler was utilized. Geweke test and Gelman-Rubin statistics, as suggested by Geweke [16] and Gelman et al. [17], show that the Markov chain converges to a stationary distribution. We used trace plots, coupling from the past plots, and sample autocorrelation plots to examine the chain's behavior, as well as to determine the burn-in period and autocorrelation lag.

It is important to decide which model provides the best fit to the dataset, the comparison of models was done using Akaike Information Criteria (AIC), Bayesian Information Criteria (BIC), Deviance Information Criteria (DIC) and Bayes factor.

6. Application in real life data

The models' applicability was tested by applying them to infectious illness data relating to kidney infection that occurred during catheter implantation [18]. It includes 38 patients' first and second recurrence times of infection from catheters used with portable dialysis equipment. For each patient in a cluster, these two times of infection are clustered together. Other pertinent data includes infection duration, patient age, gender (0 for male and 1 for female), and illness kinds such as Glomerulo Nephritis (GN), Acute Nephritis (AN), and Polycystic Kidney Disease (PKD).

To begin, the Kolmogorov–Smirnov test is used to determine the goodness of fit for kidney infection data, and the p-values obtained for the first and second recurrences are large enough to rule out the hypothesis that the first and second recurrence times follow the distributions with survival functions as given in Eqs. (11) and (14) in univariate case and it is also assumed to be appropriate for bivariate case. The corresponding p-values are given in **Table 1**. The posterior summary of the proposed models are presented in **Tables 2** and **3**. It consists of estimate (posterior mean), standard error, 95% lower and upper credible limits, GR statistics values with p-values and Geweke test values. From **Tables 2** and **3**, It is observed that we can observe that regression coefficients for all the models are more or less same. Also for all these proposed models, the value zero is not a credible value for all the credible intervals of the regression coefficients, so all the covariates are seems to be significant. To test the models' accuracy, we created 95% and 50% predictive intervals from a generated random sample based on a predictive distribution as described by Sahu et al. [15], and counted the total number of actual recurrence times for first and second kidney infections that fell within the intervals. The 95 percent and 50 percent predictive intervals are contained in the 95% and 50% predictive intervals for Models I and II, respectively, 76, 58, and 76, 60 out of 76

Distribution	Recurrence times	
	First	Second
Generalized Rayleigh	0.98078	0.99889
Exponential power	0.96291	0.75766

Table 1.
p-Values of K-S Statistics for goodness of fit test for kidney infection data set.

Parameter	Estimate	Standard error	Lower credible limit	Upper credible limit	Geweke values	p values	Gelman & Rubin values
Burn in period = 5600; autocorrelation lag = 275							
α_1	0.3716	0.0312	0.3118	0.4283	1.0007	−0.0048	0.4980
α_2	0.4253	0.0455	0.3326	0.5044	1.0003	−0.0045	0.4981
λ_1	0.0032	0.0004	0.0023	0.0041	1.0008	−0.0017	0.4992
λ_2	0.0026	0.0004	0.0018	0.0034	1.0031	−0.0052	0.4979
ξ	1.1287	0.0423	1.0722	1.2196	1.0032	−0.0095	0.4961
β_1	0.0153	0.0041	0.0083	0.0238	1.0052	−0.0042	0.4983
β_2	−1.0792	0.2740	−1.6013	−0.5350	1.0001	0.0070	0.4983
β_3	0.0021	0.0004	0.0012	0.0029	1.0008	−0.0060	0.4975
β_4	0.0031	0.0004	0.0022	0.0040	1.0005	−0.0010	0.4995
β_5	−0.2149	0.0514	-0.3031	−0.0947	1.0008	0.0059	0.5023

Table 2.
Posterior results with baseline generalized Rayleigh distribution.

Parameter	Estimate	Standard error	Lower credible limit	Upper credible limit	Geweke values	p values	Gelman & Rubin values
Burn in period = 6800; autocorrelation lag = 280							
α_1	0.4440	0.0251	0.3878	0.4905	1.0010	−0.0065	0.4973
α_2	0.5040	0.0368	0.4263	0.5689	1.0002	0.0032	0.5012
λ_1	0.3120	0.0471	0.2353	0.4021	1.0001	0.0029	0.5011
λ_2	0.2150	0.0445	0.1530	0.3166	1.0016	−0.0053	0.4978
ξ	1.2061	0.0496	1.1197	1.3013	0.9999	0.0026	0.5010
β_1	0.0001	0.0001	1.7e-05	0.0002	1.0003	0.0036	0.5014
β_2	−2.5247	0.3867	−3.2854	−1.7454	1.0021	0.0071	0.5014
β_3	0.0020	0.0004	0.0012	0.0029	1.0006	0.0119	0.5047
β_4	0.0031	0.0004	0.0021	0.0040	1.0003	0.0107	0.5042
β_5	−0.9916	0.4466	−1.8481	−0.1704	1.0027	0.0001	0.5000

Table 3.
Posterior results with baseline exponential power distribution.

observations. This demonstrates that the two models are appropriate for the data. Model-I is a better model in terms of AIC, BIC, and DIC values, since it has lower AIC, BIC, and DIC values than Model-II in **Table 4**. However, because the difference between AIC, BIC, and DIC values for Model I and Model II is so little, AIC, BIC, and DIC values are not suitable for deciding between the two models. To compare model u with model v, we use the Bayes factor (**Table 5**). Model-I is better than Model-II, since the equivalent value of $2\,log\,(B_{uv})$ is larger than 10, suggesting that there is a very strong positive to favor Model-I over Model-II for the provided dataset, confirming our earlier conclusion in **Table 4**. As a result of all of the demonstrated comparison criteria, we can conclude that Model-I is superior to Model-II in terms of modeling kidney infection data.

Model no.	AIC	BIC	DIC
Model I	638.5262	654.9020	625.1190
Model II	700.3005	716.6763	686.8069
Model III	691.5817	706.3200	720.7843
Model IV	702.0827	716.8210	689.8978

Table 4.
AIC, BIC and DIC values for all models.

Numerator model against denominator model	$2log_e(B_{uv})$	Range	Evidence against model in denominator
$Model - I$ against $Model - II$	63.23936	>10	Very Strong Positive

Table 5.
Bayes factor values and decision for test of significance for frailty fitted to kidney infection data set.

7. Discussion

In this study, we examined a new Lindley shared frailty model under generalized Rayleigh and exponential power as baseline distributions.

To suit all of the proposed models, the Metropolis-Hastings and Gibbs sampler was used. The proposed models were used to assess kidney infection data, and the best model was suggested. To conduct the analysis, we used self-composed programs in the R statistical software.

All of the exhibited comparison criteria indicated that the Lindley shared frailty model with generalized Rayleigh baseline distribution is superior to exponential power baseline distribution and without frailty models for modeling kidney infection data under the identical baseline distributions. The estimates of frailty variance are 0.9415 and 0.9739, which are high in all the proposed models indicating that there is a strong evidence of a high degree of heterogeneity among the patients in the population. A few patients are anticipated to be exceptionally inclined to infection compared to others with the same covariate values. Some patients are expected to be very prone to infection compared to others with the same covariate values. Also we can say that there is a strong positive correlation between the two infection times for the same patient.

The most important properties of the proposed models that were not mentioned in the previous study are the estimates of the frailty variances are high in all proposed models as compared to previous study given by McGilchrist and Aisbett [18] on log-normal frailty, Hanagal and Bhambure [19], the disease type GN and AN has lower infection rates as compared to other covariates. All the covariates are significant factors for kidney infection, but the disease type are insignificant in the previous proposed frailty models (see [4]). It is very crucial to be mention that Lindly shared frailty model based on generalized Rayleigh baseline distribution is performed better to analyze kidney infection data than other frailty models [4, 19].

Author details

Arvind Pandey[1] and Lalpawimawha[2*]

1 Department of Statistics, Central University of Rajasthan, Rajasthan, India

2 Department of Statistics, Pachhunga University College, Mizoram, India

*Address all correspondence to: raltelalpawimawha08@gmail.com

References

[1] Vaupel, J. W., Manton, K. G., Stallard, E. 1979. The impact of heterogeneity in individual frailty on the dynamics of mortality. Demography, 16(3), 439-454.

[2] Gupta, R.D., and Kundu, D. (2001). Exponentiated Exponential Family: An Alternative to Gamma and Weibull Distributions, Biometrical Journal,43 (1), 117–130.

[3] Kheiri, S., A. Kimber, and M.R. Meshkani. 2007. Bayesian analysis of an inverse Gaussian correlated frailty model. Computational Statistics and Data Analysis 51: 5317-5326.

[4] Hanagal, D.D., Dabade, A.D. 2013. Compound negative binomial shared frailty models for bivariate survival data. Statistics and Probability Letters, 83, 2507-2515.

[5] Pandey, A., Bhushan, S., Lalpawimawha, R. 2018. Shared frailty models with baseline generalized Pareto distribution, Communications in Statistics-Theory and Methods, DOI: 10.1080/03610926.2018.1500597

[6] Pandey, A., Bhushan, S., Lalpawimawha, R., Misra, P.K. 2019. Comparison of additive shared frailty models under Lindley baseline distribution, Communications in Statistics-Simulation and Computation, DOI: 10.1080/ 03610918.2019.1664573

[7] Lindley, D.V. 1958. Fiducial distributions and Bayes' theorem. Journal of the Royal Statistical Society, Series B, 20, 102-107.

[8] Ghitany, M.E., Atieh, B., Nadarajah, S. 2008. Lindley Distribution and Its Applications. Mathematics and Computers in Simulation, 78(4), 2008, 493-506.

[9] Ghitany, M.E., Alqallaf, F., Al-Mutairi, D.K., Hussain, H. 2011. A Two Parameter Weighted Lindley Distribu-tion and Its Applications to Survival Data. Mathematics and Computers in Simulation, 81(6), 1190-1201.

[10] Bakouch, H.S., Al-Zahrani, B.M., Al-Sho, A.A., Marchi, A.A., Louzada, F. 2012. An Extended Lindley Distribution. Journal of the Korean Statistical Society, 41(1), 75-85.

[11] Ghitany, M., Al-Mutairi, D., Balakrishnan, N. and Al-Enezi, I., 2013. Power Lindley distribution and associated inference. Computational Statistics and Data Analysis, 64, 20-33.

[12] Shanker, R., Sharma, S., Shanker, R. 2013. A Two-Parameter Lindley Distribution for Modeling Waiting and Survival Times Data. Applied Mathematics, 4, 363-368.

[13] Surles, J. G., Padgett, W.J. (2001). Inference for reliability and stress-strength for a scaled Burr Type X distribution. Lifetime Data Anal. 7,187-200.

[14] Ibrahim, J.G., Chen, Ming-Hui, Sinha, D., 2001. Bayesian Survival Analysis. Springer-Verlag.

[15] Sahu, S.K., D.K. Dey, H. Aslanidou, & D. Sinha. 1997. A Weibull regression model with gamma frailties for multivariate survival data. Life time data analysis, 3, 123-137.

[16] Geweke, J. 1992. "Evaluating the Accuracy of Sampling-Based Approaches to the Calculation of Posterior Moments." In Bayesian Statistics 4 (eds. J.M. Bernardo, J. Berger, A.P. Dawid and A.F.M. Smith), 169-193. Oxford: Oxford University Press.

[17] Gelman, A., & D.B. Rubin. 1992. A single series from the Gibbs sampler provides a false sense of security. In Bayesian Statistics 4 (J. M. Bernardo, J. 0.Berger, A. P. Dawid and A. F. M. Smith, eds.). Oxford: Oxford Univ. Press, 625-632.

[18] McGilchrist, C.A., & C.W. Aisbett. 1991. Regression with frailty in survival analysis. Biometrics, 47, 461-466.

[19] Hanagal, D.D., & Bhambure, S.M. (2016). Modeling bivariate survival data using shared inverse Gaussian frailty model. Communications in Statistics-Theory and Methods, 45(17), 4969-4987.

Section 3

Other Statistical Techniques

Chapter 7

Network Meta-Analysis Using R for Diabetes Data

Nilgün Yildiz

Abstract

The objective of a meta-analysis is usually to estimate the overall treatment effect and make inferences about the difference between the effects of the two treatments. Meta-analysis is a quantitative method commonly used to combine the results of multiple studies in the medical and social sciences. There are three common types of meta-analysis. Pairwise, Multivariate and Network Meta-analysis. In general, network meta-analysis (NMA) offers the advantage of enabling the combined assessment of more than two treatments. Statistical approaches to NMA are largely classified as frequentist and Bayesian frameworks Because part of NMA has indirect, multiple comparisons, As reports of network meta-analysis become more common, it is essential to introduce the approach to readers and to provide guidance as to how to interpret the results. In this chapter, the terms used in NMA are defined, relevant statistical concepts are summarized, and the NMA analytic process based on the frequentist and Bayesian framework is illustrated using the R program and an example of a network involving diabetes treatments. The aim of the article is to compare the basic concepts and analyzes of network meta-analysis using diabetes data and the treatment methods used.

Keywords: Network meta-analysis, fixed effect model, random-effects model, forest plot, network graph, direct evidence plot

1. Introduction

Meta-analysis is used to synthesize the results of more than one study and overall effect size is considered to be valid only when some required assumptions are satisfied [1]. An increasing number of options for alternative medical treatment has given rise to the need for comparative effectiveness research [2, 3]. A randomized, controlled trials used to compare different treatment options are generally seen to be infeasible, there is a need for other methodological approaches. Since it makes it possible to combine data from many different studies so that a total estimate of treatment effect can be provided, a meta-analysis integrated into a systematic review is generally seen to be a useful statistical tool. On the other hand, there is an important limitation of standard meta-analysis; only two interventions can be compared at a time. When you have several treatment options to capitalize on, only partial information can be provided by a series of individual meta-analysis since only the questions about pairs of treatments can be answered in this way, which leads to difficulties in making optimal clinical decisions since each meta-analysis is just one constituent of the whole picture.

There is an increasing need for a method to be used to summarize evidence across many interventions [4]. In order to assess a number of interventions in terms of their relative effectiveness and to synthesize evidence from a set of randomized trials, network meta-analysis (or multiple treatments meta-analysis or mixed-treatment comparison) was created [5–7]. This method is built on the analysis of direct evidence (coming from research that directly randomizes treatments of interest) and indirect evidence (coming from research that compares treatments of interest with a common comparator) [8]. The benefits incurred by network analysis becoming increasingly popular have been reported in some applications and methodological articles [2, 9, 10]. **Figure 1** shows the number of network meta-analysis (NMA) studies that have been published.

Despite the fact that network meta-analysis shares many underlying assumptions with pairwise meta-analysis, it is not so much accepted as pairwise meta-analysis and thus criticized more [11].

The assumptions required by NMA about similarity, transitivity, and consistency [12–17] are methodologically, logically, and statistically more strict [18, 19] because it should be examined whether each of these is satisfied or not [15, 20, 21].

For NMA, there are some methods to calculate the contribution of direct (and indirect) evidence of each comparison to its own NMA estimate, but how to define the contribution of each study to another estimate of treatment effect is an issue of greater ambiguity. There are a number of proposals made in the literature, each of which is based on a different approach but many of them are not without limitations and generally, there are contradictions between their results [20, 22–24]. There are some investigations having been conducted on the proportions of direct and indirect evidence in the past. One of these is the method of "back calculation" [21] introduced by Dias and some others have been proposed within a Bayesian framework [25]. There is even one proposed within a frequentist context [13]. In inverse variance method-based NMA, NMA estimates refer to linear combinations of treatment effect estimates from primary studies having coefficients that make up the rows of the hat matrix. It is easy to obtain the direct evidence proportion of a study or a comparison from the diagonal elements that the respective hat matrix has [13]. Dias and others proposed "node splitting" as an alternative. Node splitting refers to the estimation of the indirect evidence for comparison by modeling out all studies providing direct information for this comparison [25]. Additions were made to this method [26] and called "side splitting" by others [9]. There are different

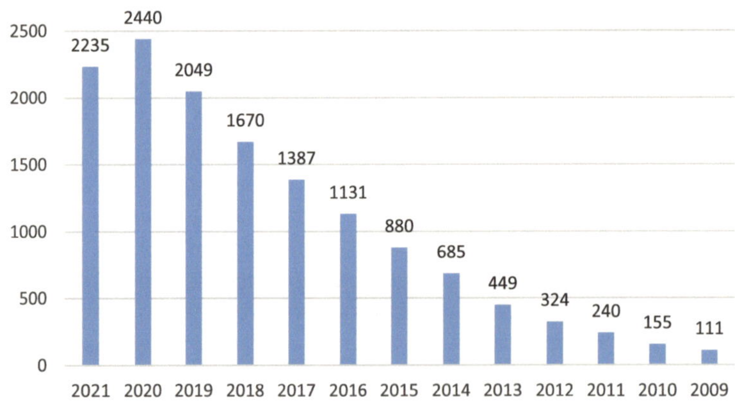

Figure 1.
Number of network meta-analysis publications (search PubMed until January 2020).

interpretations of the term "side" in the literature; for example, it was interpreted as an edge in the network graph by White [9] while it was interpreted as SIDE, an abbreviation of "Separating Indirect and Direct Evidence" by others [27, 28]. There is another way of quantifying the indirect evidence proposed by Noma and others, including the factorization of the total likelihood into separate component likelihoods [14]. Yet, none of these authors have attempted to make a definition or estimation of the contribution of each study to a given comparison in the network.

There are six basic steps that every NMA should follow, regardless of the analytic model chosen. These steps include

1. Understand network geometry,

2. Understand key concepts and assumptions,

3. Conduct analysis and present results,

4. Examine model assumptions through local and global tests,

5. Create a hierarchy of competing interventions (ranking),

6. Conduct heterogeneity and sensitivity analyses.

The network plot is fundamental to an NMA because it helps visualize the available studies and few of evidence across the multiple comparisons. In such a plot, each treatment/comparator identified in the review is represented by a node, and direct evidence comparing two interventions (i.e., studies which directly compared these two interventions) are represented via edges, connecting the respective nodes. The network plot of our example is presented in **Figure 2**.

In **Figure 2**, a network of treatments for type 2 diabetes is shown. The function served by the lines between the treatment nodes is to show which comparisons have been made in trials that are randomized. The absence of a line between two nodes means that there are no studies (that is, no direct evidence) comparing the two

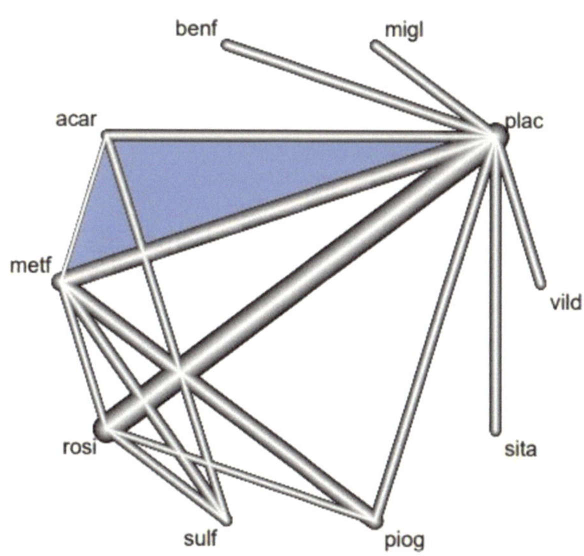

Figure 2.
A graph of the network generated by using the net graph function for the diabetes data.

drugs. A network meta-analysis refers to an analysis of the data from all of these randomized trials at the same time. By means of a network meta-analysis, it is possible to estimate the relative effectiveness of two treatments even if they are not compared by any studies. For example, no comparison has been made between rosi and acar in any study but by using a common comparator (placebo), an indirect comparison can be made between them. After denoting rosi, acar, and placebo as treatments A, B, and C, respectively, it is possible to have an indirect comparison (AB) by subtracting the meta-analytic estimates of all studies of acar versus placebo (BC) from the estimate of all studies of rosi versus placebo (AC): AB indirect meta-analysis _ AC direct meta-analysis _ BC direct meta-analysis. If there is direct evidence (such as metf vs. sulf in (**Figure 2**), direct and indirect estimates can be combined by the network meta-analysis and mixed effect size can be calculated as the weighted average of the direct evidence (studies comparing metf and sulf directly) and the indirect evidence (for example, studies comparing metf and acar via placebo). The network constructed by studies of metf versus acar, metf versus placebo, and acar versus placebo is often named as a loop of evidence. By using indirect estimates, information can be provided on comparisons for which there are no trials. In this way, the accuracy of the direct estimate can be enhanced through the reduction of the width of the CIs in comparison with the direct evidence alone [9].

In a network meta-analysis, all the direct and indirect evidence can be utilized. Empirical studies have concluded that compared to a single direct or indirect estimate, it can produce more precise estimates of the intervention effects [2, 29]. Moreover, network meta-analysis has the potential of yielding data for comparisons made between pairs of interventions having never been evaluated within individual randomized trials. The comparison of all interventions of interest simultaneously in the same analysis makes it possible to estimate their ranking relatively for a given result. The purpose of this study is to show how analysis can be done with the network meta-analysis method using the R package program. Network meta-analysis as a functional method. It is to show that it can be done flexibly and easily with the R program to help researchers interested in this subject.

This chapter is organized as follows, In the next sections, we present a review of the methods for NMA as identified in our literature search. In Section 2, we present key concepts and the basic methodology for NMA. In Section 3 Diabetes treatments data is used as an example. The last section presents conclusions about our research and results found by network meta-analysis of diabetes data using the R program.

2. Conceptual issues and underlying assumptions in network meta-analysis

There may be different alternatives for the treatment of the same health condition and what makes NMA special is that through the synthesis of direct and indirect estimates for their relative effects, it allows the selection of the best treatment. Head-to-head studies can be conducted to directly compare two treatments A and B (AB studies). It is also possible to get an indirect estimate from studies in which these two treatments are compared with a common comparator treatment C, namely, AC and BC studies [9]. If we have both direct and indirect estimates, then we can combine them to estimate a mixed-treatment effect, as you can see in the left panel of **Figure 3**. In practice, there are numerous interventions for most health conditions that have been compared in various randomized trials and build a network of evidence. For the comparison of treatments within such a network, there may be direct and many different indirect estimates obtained through many different comparators, as illustrated in the example in the right panel of **Figure 2**.

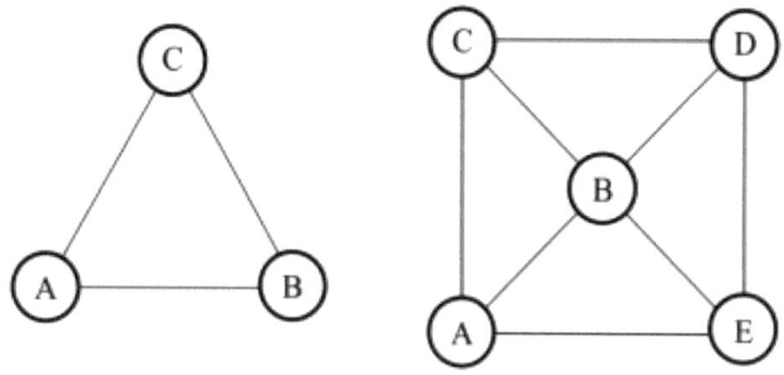

Figure 3.
Each circle represents an intervention, and lines represent direct comparisons.

Using NMA, all these different pieces of information can be compared so that an internally consistent overall estimate of the relative effects of all treatments can be produced. Researchers are still disputing about how valid it is to use indirect treatment comparisons (indirect evidence) while making decisions. There are strong arguments against using such evidence especially when there are direct treatment comparisons (direct evidence) [11, 30–32]. A focus of criticism is the nature of the evidence provided by NMA. Although patients in a randomized clinical trial (RCT) are randomly assigned to each of the treatments that are compared, it cannot be argued that the treatments are randomized across the included trials.

Thus, indirect comparisons can be defined as non-randomized comparisons and correspondingly they provide observational evidence rather than randomized evidence. Consequently, indirect treatment comparisons may be more susceptible to biased treatment effect estimates, due, for example, to confounding (for example, when randomized AB and AC studies are systematically different from BC; [2] and selection bias (e.g., when the selection of comparator in the study is based on the relative treatment effect [33].

2.1 Indirect comparisons

Consider trial 1, a two-arm trial of the comparison "B–A", and trial 2, a two-arm trial of the comparison "C–B". If the estimated effect sizes in these trials are $\hat{\delta}_1^{AB}$ in trial 1 and $\hat{\delta}_2^{BC}$ in trial 2, then an indirect comparison of "C–A" may be obtained as $\hat{\delta}_{indirect}^{AC} = \hat{\delta}_1^{AB} + \hat{\delta}_2^{BC}$.

Through indirect comparison, the benefits of randomization can be maintained in each trial, and differences across the trials are allowed (e.g.,, in baseline risk) if only the prognosis of the participants but not their response to treatment is affected by these differences (in whichever metric is chosen as a measure of effect size). However, the indirect comparison is based on the assumption that the treatment named as B is the same in both trials so that its effects are nullified when "B-A" and "C-B" are added together. It is not possible to test whether the difference between A and C is truly reflected by an indirect comparison without having further information. The comparison of the indirect comparison with a direct comparison would be allowed by a third trial of "C–A" (yielding result $\hat{\delta}_3^{AC}$). The network of these three trials can be said to be consistent only if the underlying treatment effects are related to each other as follows:

$$\hat{\delta}_3^{AC} = \hat{\delta}_1^{AB} + \hat{\delta}_2^{BC} \tag{1}$$

Here $\hat{\delta}_1^{AB}$, $\hat{\delta}_2^{BC}$ and $\hat{\delta}_3^{AC}$ represent the actual effects that underlay the three studies. In practice, it is not very likely for Eq. (1) to hold for a particular set of three trials such as the ones described earlier. The reason behind this may be discussed either in terms of heterogeneity (because, within each treatment comparison, each individual study may not fully represent all studies in this particular comparison) or in terms of inconsistency (because, across treatment comparisons, important differences in the types of studies contributing to the comparisons exist). We will give more detailed information on these two concepts in subsequent sections [34].

2.2 Heterogeneity

The existing research has widely investigated heterogeneity in meta-analysis, referring to the situation where multiple studies focused on the same research question have different underlying values regarding the effect measure that is being estimated. The way of understanding heterogeneity in the network meta-analysis scenario is to keep the treatment comparison constant while changing the study index. In particular, the existence of heterogeneity can be argued for comparison 'B–A' if $\hat{\delta}_i^{AB} \neq \hat{\delta}_j^{AB}$ for some pair of studies i and j. It has been claimed that heterogeneity is an inevitable part of a meta-analysis [35] indicating that it is not likely that two trials of the same pairwise comparison are to have equal underlying treatment effects. Hence, within the context of Eq. (1), it is unlikely that the equality holds since the particular instance of "C–A", which is examined in trial 3, will probably not represent all instances of "C–A" comparisons (and this holds true for trials 1 and 2 for their respective treatment comparisons). A random-effects model is a common way of allowing for heterogeneity. This assumes that the main effects in multiple runs of the same comparison arise from a common distribution, usually a normal distribution; namely,

$$\delta_i^{JK} \sim N\left(\delta^{JK}, \tau_{JK}^2\right) \tag{2}$$

for pairwise comparison JK (taking values AB, AC, or BC in the running example) [34].

2.3 Consistency

Consistency is the statistical manifestation of transitivity [12]. An additional way of making implicit inferences about the plausibility of the transitivity assumption is to check the network for consistency. What is meant by consistency is the statistical agreement between observed direct and (possibly many) indirect sources of evidence. A simple network can only contain treatments A, B, and C.

A consistency equation is generally used to express the relationship that is desirable between direct and indirect sources of evidence for a single comparison

$$\delta^{AC} = \delta^{AB} + \delta^{BC} \tag{3}$$

where the mean effect size across all studies of comparison JK is represented by JK. (Under a fixed-effect meta-analysis model where the absence of heterogeneity is assumed, dJK represents a fixed (common) treatment effect for comparison JK). We refer to evidence that satisfies the consistency equation as showing consistency. We show this in **Figure 4(a)** as a three (non-touching) solid-edge relationship triangle in a network with only two-arm trials. Each edge represents one or more

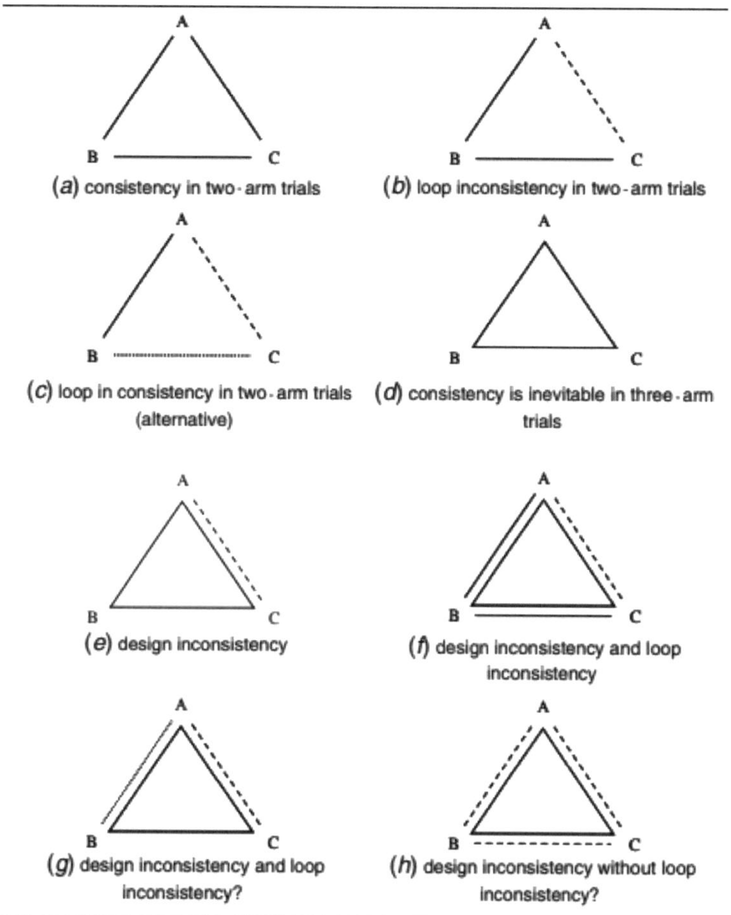

Figure 4.
Graphical representation of consistency, loop inconsistency and design inconsistency.

two-arm trials that compare two treatments identified at either end of the edge. Using the same line style (a solid line), we draw all three edges to describe the situation where there is no contradiction (inconsistency) between them, that is, Eq. (2) is valid [34].

2.4 Loop inconsistency

When studies focused on various treatment comparisons are highly different in such a way that their effect sizes are affected, the consistency Eq. (2) might not be valid; thus, the effect sizes are not "added up" around the loop in the figure. This is called loop inconsistency and is shown by drawing edges using different line styles (**Figure 4(b)**). Loop inconsistency may only result from when there are different comparisons made in at least three separate study groups (e.g., studies "B–A", "C–A" and "C–B"). Equivalently, it can only occur when we have both indirect and direct estimates of effect size (e.g., when "C–B" is measured both directly and through "A" indirectly) [34]. Some examples showing the causes of loop consistency are given below:

2.5 Multi-arm trials

Generally, some studies having more than two treatment arms are included in a network meta-analysis. In fact, about a quarter of randomized trials involve more than two arms [36], so it is important to select appropriate methods while dealing with the condition.

When there is the presence of multi-arm trials in an evidence network, the definition of loop inconsistency becomes more complicated. It is not possible to loop inconsistency in a multi-arm trial. As a result, consistency can occur for a network either structurally (because all studies include all treatments) or through observation (when assumptions about equality of direct and various indirect comparisons hold across studies), or by means of a combination of the two.

Also, loop inconsistency cannot be properly defined using Eq. (2) anymore, since average effect sizes, δ^{JK}, refer to pairwise comparisons made from a combination of possibly inconsistent loops (e.g., from the two-arm trial) and naturally consistent loops (i.e., from multi-arm trials). In our drawings, multi-arm trials are shown using a closed (merged) polygon (**Figure 4(d)**) [34].

2.6 Transitivity

The purpose of an NMA is to improve the decision-making process for making choices between alternative treatments for a specific health condition and a target population. Hence, the estimates intended to be estimated in an NMA are the mean relative treatment effect sizes among the treatments competing with each other as they are expected to be present in the target population. If unbiased estimates are yielded by studies involved in the dataset and if a representative sample of the population addressed is constituted by these studies, then estimates generated by an NMA model for these parameters will be unbiased and consistent. The same set of assumptions is adopted by NMA as a pairwise meta-analysis [37], but there is also another assumption adopted by it which can be difficult to assess [38] and is called transitivity [39], (also called similarity [40, 41], or exchangeability [42]). Transitivity means that information for comparison between treatments A and B can be attained through another treatment C using comparisons A to C and B to C. It is not possible to test his assumption statistically, but it is possible to evaluate its validity in a conceptual and epidemiological way [21].

What is meant by the transitivity assumption is that direct evidence from studies AC and BC can be combined to gain insights (indirectly) about AB comparison. However, this will be open to questioning if there are significant differences in the distribution of effect modifiers (variables or characteristics that alter the observed relative effects, e.g., the mean age of participants and treatment dose) across the AC and BC trials, which yield insights about the indirect comparison [24, 39]. An effect modifier might have different effects across studies of the same comparison (e.g., the mean age of participants may differ across AC trials), but if its distribution across comparisons (AC and BC) is similar, the assumption of transitivity may still hold [21]. As a consequence, how plausible the transitivity assumption is can be assessed by reviewing the collection of studies for significant differences in the distribution of effect modifiers. Assuming that the studies are similar, the assumption of transitivity may be realistic, on the condition that there aren't any unknown modifiers of the relative treatment effect [43]. It is clear that such an assessment of transitivity may not be possible when the effect modifiers are not reported or when the number of studies per treatment comparison is low [12]. If there are significant differences identified and sufficient data is available, the transitivity of the network can be enhanced by using a network meta-regression. This might indicate, for example, that it is necessary for the common comparator treatment C to be similar in the AC and BC studies in terms of dose, modes of administration, duration, etc.

In an NMA of studies conducted to compare fluoride treatments administered to prevent dental caries, the definition of placebo differed between fluoride toothpaste studies and fluoride rinse studies [44], casting doubt on how plausible the transitivity assumption is and thus challenging the reliability of the NMA results. In another example, Julious and Wang [45] focused on how the use of placebo as an intermediate comparator can result in the distortion of the results of indirect comparisons due to changes in the population's placebo response over the years; for instance, there might be a bias in the indirect estimate for A versus B when studies that compare treatment A versus placebo are older than studies that compare B versus placebo. Other ways used to formulate the transitivity assumption is to suppose that the true relative effect of A versus B is the same in the fixed-effects model or may vary across studies in the random-effects model, regardless of the treatments compared in each study [42, 46], that "missing" treatments in each trial are randomly missing [5] or, equivalently, that the choice of treatment comparisons in trials is not related directly or indirectly to the relative efficacy of the interventions. Finally, arguing that the patients included can be randomly distributed to any of the treatments in the network is an alternative way of postulating this assumption [21].

However, this does not mean that the assumption of transitivity will necessarily be valid. It should be stated that the absence of statistical inconsistency does not offer any evidence to prove the validity of the transitivity assumption that is essentially an assumption that cannot be tested as discussed in the previous section. Therefore, the conduct of an NMA should be preceded by a conceptual/theoretical evaluation of the transitivity assumption besides statistical tests for inconsistency [12] and the studies that are included in an NMA should always be reviewed for important differences that can be seen in patients, interventions, outcomes, study design, methodological characteristics, and reporting biases [2, 9, 14, 32, 43].

2.7 Design inconsistency

What is meant with the "design" of a study is a set of treatments that are compared within the study, recognizing that it is different from traditional interpretations made for the term. Then, differences in effect sizes among studies including different sets of treatments are referred to by design inconsistency. While allowing for this variation, it is implicitly assumed that different designs (i.e., different treatment sets included) can serve the function of a proxy for one or more important modifiers of effect [47]. Design inconsistency is depicted in **Figure 4(e)**, in which different line styles represent possible contradictions between study designs. The AC effect size depicted with a solid line in the three-arm trial is different from the AC effect size in the two-arm trial depicted with a dashed line. It is possible to see design inconsistency as a special case of heterogeneity since study designs correspond to a study-level covariate that has the potential to change effect sizes in the study, as can occur in a standard meta-regression analysis. It should be noted that in a network of only two-arm studies, additional insights provided by loop inconsistency cannot be provided by the concept of design inconsistency. In the case of a multi-arm trial, loop inconsistency in two-arm trials means design inconsistency (**Figure 4(f)**). The reason for this is that the multi-arm trial must be self-consistent, so the effect sizes of the multi-arm trial should be different from those of at least one of the two-arm trials: our definition of design inconsistency. Nevertheless, what is implied by design inconsistency for loop inconsistency is less clear. Design consistency with one three-arm trial and two two-arm trials is shown in **Figure 4(g)**. It is possible to create a loop by subtracting the pairwise BC comparison from the three-arm trial and then by comparing it to the two-arm trials. But, in this way, the existence of a consistent loop in the three-arm experiment is overlooked and thus it is unclear whether this network should be defined as exhibiting loop inconsistency. Also, it is seen in **Figure 4(h)** that the two-arm trials are consistent among themselves, but the effect sizes are different from the effect sizes of the multi-arm trial. Does this show design inconsistency without loop inconsistency? [34].

2.8 Similarity

In order to make a comparison among the clinical trial studies used for analysis, it must be assumed that there is a similarity in the methodology used in the studies [12, 44]. The assessment of similarity is qualitatively performed on each of the selected articles from a methodological point of view and is not a hypothesis that can be tested statistically. The technique used to investigate similarity is the population, intervention, comparison, and outcome (PICO) technique [17]. Examination of similarity among the studies used for analysis is based on the following four items: clinical characteristics of study subjects, treatment interventions, comparison treatments, and outcome measures. In cases where the similarity assumption is not satisfied, the other two assumptions are also negatively affected [24] and moreover, there is also a need to check for the heterogeneity error [18, 21].

2.8.1 Network diagrams

One way of graphically depicting the structure of a network of interventions is a network diagram [12]. Such a graph is comprised of nodes that represent the interventions in the network and lines that show the available direct comparisons between pairs of interventions. An example of a network diagram including four interventions is given in **Figure 3**. In this example, in order to show the presence of a three-arm study, distinct lines that form a closed triangular loop have been added. It should be noted that complex and useless network diagrams may be yielded by such presentation of multi-arm studies; in this case, a tabular format can be preferred to depict multi-arm studies (**Figure 5**).

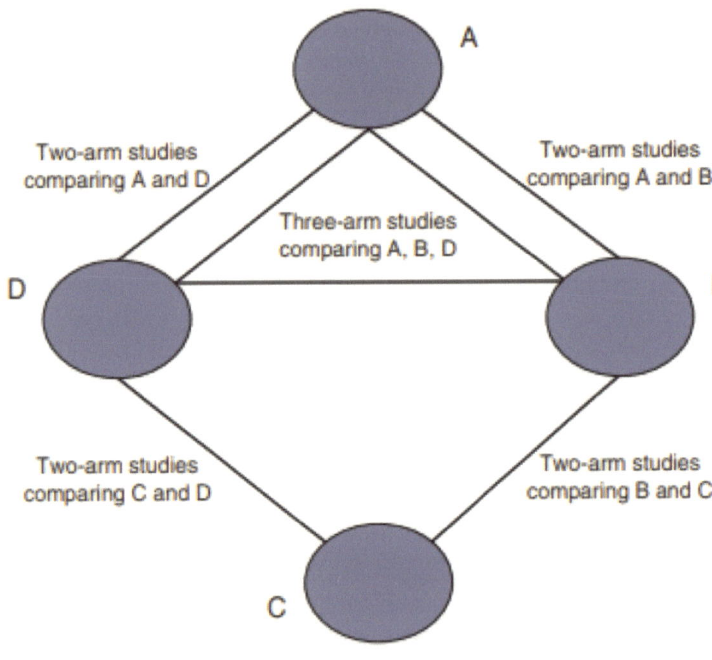

Figure 5.
Example of network diagram with four competing interventions and information on the presence of multi-arm randomized trials.

3. Illustrating example

The estimation of the relative effects on HbA1c change, of adding different oral glucose-lowering agents to a baseline sulfonylurea therapy in patients with type 2 diabetes, was the aim of the network meta-analysis in Diabetes. Systemic literature research was carried out on all relevant articles that were published from January 1993 to June 2009 in Medline and Embase. The search strategy was restricted to "randomized controlled 170 Statistical Methods in Medical Research 22(2) trials", "sulfonylurea or sulphonylurea" and "humans". This initial search was confirmed by combining each of the Medical Subject Headings key words "chlopropamide", "glibenclamide", "glyburide", "gliclazide", "glimepiride", "glipizide", "gliquidone", "tolbutamide" on the one hand and 'RCT' on the other hand. No language restriction was applied. R program was used to analyze the data (**Figure 6**).

An original dataset offered by Senn [48] will be used in our first network meta-analysis. In this dataset, there are effect size data obtained from randomized controlled trials that compare different medications for diabetes. The effect size obtained for all comparisons represents the mean difference (MD) of diabetic patients' HbA1c value in the posttest. What is represented by this value is the concentration of glucose found in the blood, which is aimed to be decreased with diabetic medication. As can be seen, there are 28 rows that represent the treatment comparisons and seven columns in the data. In the first column, TE, there is the effect size of each comparison, and the respective standard error is contained in se TE. In case effect size data that have already been calculated for each comparison might not be possessed.

	TE	seTE	treat1.long	treat2.long	treat1	treat2	studlab
1	-1.90	0.1414	Metformin	Placebo	metf	plac	DeFronzo1995
2	-0.82	0.0992	Metformin	Placebo	metf	plac	Lewin2007
3	-0.20	0.3579	Metformin	Acarbose	metf	acar	Willms1999
4	-1.34	0.1435	Rosiglitazone	Placebo	rosi	plac	Davidson2007
5	-1.10	0.1141	Rosiglitazone	Placebo	rosi	plac	Wolffenbuttel1999
6	-1.30	0.1268	Pioglitazone	Placebo	piog	plac	Kipnes2001
7	-0.77	0.1078	Rosiglitazone	Placebo	rosi	plac	Kerenyi2004
8	0.16	0.0849	Pioglitazone	Metformin	piog	metf	Hanefeld2004
9	0.10	0.1831	Pioglitazone	Rosiglitazone	piog	rosi	Derosa2004
10	-1.30	0.1014	Rosiglitazone	Placebo	rosi	plac	Baksi2004
11	-1.09	0.2263	Rosiglitazone	Placebo	rosi	plac	Rosenstock2008
12	-1.50	0.1624	Rosiglitazone	Placebo	rosi	plac	Zhu2003
13	-0.14	0.2239	Rosiglitazone	Metformin	rosi	metf	Yang2003
14	-1.20	0.1436	Rosiglitazone	Sulfonylurea	rosi	sulf	Vongthavaravat2002
15	-0.40	0.1549	Acarbose	Sulfonylurea	acar	sulf	Oyama2008
16	-0.80	0.1432	Acarbose	Placebo	acar	plac	Costa1997
17	-0.57	0.1291	Sitagliptin	Placebo	sita	plac	Hermansen2007
18	-0.70	0.1273	Vildagliptin	Placebo	vild	plac	Garber2008
19	-0.37	0.1184	Metformin	Sulfonylurea	metf	sulf	Alex1998
20	-0.74	0.1839	Miglitol	Placebo	migl	plac	Johnston1994
21	-1.41	0.2235	Miglitol	Placebo	migl	plac	Johnston1998a
22	0.00	0.2339	Rosiglitazone	Metformin	rosi	metf	Kim2007
23	-0.68	0.2828	Miglitol	Placebo	migl	plac	Johnston1998b
24	-0.40	0.4356	Metformin	Placebo	metf	plac	Gonzalez-Ortiz2004
25	-0.23	0.3467	Benfluorex	Placebo	benf	plac	Stucci1996
26	-1.01	0.1366	Benfluorex	Placebo	benf	plac	Moulin2006
27	-1.20	0.3758	Metformin	Placebo	metf	plac	Willms1999
28	-1.00	0.4669	Acarbose	Placebo	acar	plac	Willms1999

Figure 6.
Diabetes example and view the data.

The two treatments that are compared are represented by treat1. long, treat2. long, treat1, and treat2. As a shortened name of the original treatment name is contained in the variables treat1 and treat2, they are redundant.

We can now move forward by fitting our initial network meta-analysis model using the net metafunction. Now, we can look at the results of our first model, for now assuming a fixed-effects model.

```
Results (fixed effects model):

                    treat1 treat2      MD              95%-CI     Q leverage
DeFronzo1995          metf   plac -1.1141 [-1.2309; -0.9973] 30.89     0.18
Lewin2007             metf   plac -1.1141 [-1.2309; -0.9973]  8.79     0.36
Willms1999            acar   metf  0.2867 [ 0.0622;  0.5113]  0.05        .
Davidson2007          plac   rosi  1.2018 [ 1.1084;  1.2953]  0.93     0.11
Wolffenbuttel1999     plac   rosi  1.2018 [ 1.1084;  1.2953]  0.80     0.17
Kipnes2001            piog   plac -1.0664 [-1.2151; -0.9178]  3.39     0.36
Kerenyi2004           plac   rosi  1.2018 [ 1.1084;  1.2953] 16.05     0.20
Hanefeld2004          metf   piog -0.0477 [-0.1845;  0.0891]  1.75     0.68
Derosa2004            piog   rosi  0.1354 [-0.0249;  0.2957]  0.04     0.20
Baksi2004             plac   rosi  1.2018 [ 1.1084;  1.2953]  0.94     0.22
Rosenstock2008        plac   rosi  1.2018 [ 1.1084;  1.2953]  0.24     0.04
Zhu2003               plac   rosi  1.2018 [ 1.1084;  1.2953]  3.37     0.09
Yang2003              metf   rosi  0.0877 [-0.0449;  0.2203]  0.05     0.09
Vongthavaravat2002    rosi   sulf -0.7623 [-0.9427; -0.5820]  9.29     0.41
Oyama2008             acar   sulf -0.3879 [-0.6095; -0.1662]  0.01     0.53
Costa1997             acar   plac -0.8274 [-1.0401; -0.6147]  0.04     0.57
Hermansen2007         plac   sita  0.5700 [ 0.3170;  0.8230]  0.00     1.00
Garber2008            plac   vild  0.7000 [ 0.4505;  0.9495]  0.00     1.00
Alex1998              metf   sulf -0.6746 [-0.8482; -0.5011]  6.62     0.56
Johnston1994          migl   plac -0.9439 [-1.1927; -0.6952]  1.23     0.48
Johnston1998a         migl   plac -0.9439 [-1.1927; -0.6952]  4.35     0.32
Kim2007               metf   rosi  0.0877 [-0.0449;  0.2203]  0.14     0.08
Johnston1998b         migl   plac -0.9439 [-1.1927; -0.6952]  0.87     0.20
Gonzalez-Ortiz2004    metf   plac -1.1141 [-1.2309; -0.9973]  2.69     0.02
Stucci1996            benf   plac -0.9052 [-1.1543; -0.6561]  3.79     0.13
Moulin2006            benf   plac -0.9052 [-1.1543; -0.6561]  0.59     0.87
Lewin2007             metf   plac -1.1141 [-1.2309; -0.9973]  8.79     0.36
Willms1999            acar   metf  0.2867 [ 0.0622;  0.5113]  0.05        .
Davidson2007          plac   rosi  1.2018 [ 1.1084;  1.2953]  0.93     0.11
Wolffenbuttel1999     plac   rosi  1.2018 [ 1.1084;  1.2953]  0.80     0.17
Kipnes2001            piog   plac -1.0664 [-1.2151; -0.9178]  3.39     0.36
Kerenyi2004           plac   rosi  1.2018 [ 1.1084;  1.2953] 16.05     0.20
Hanefeld2004          metf   piog -0.0477 [-0.1845;  0.0891]  1.75     0.68
Derosa2004            piog   rosi  0.1354 [-0.0249;  0.2957]  0.04     0.20
Baksi2004             plac   rosi  1.2018 [ 1.1084;  1.2953]  0.94     0.22
Rosenstock2008        plac   rosi  1.2018 [ 1.1084;  1.2953]  0.24     0.04
Zhu2003               plac   rosi  1.2018 [ 1.1084;  1.2953]  3.37     0.09
Yang2003              metf   rosi  0.0877 [-0.0449;  0.2203]  0.05     0.09
Vongthavaravat2002    rosi   sulf -0.7623 [-0.9427; -0.5820]  9.29     0.41
Oyama2008             acar   sulf -0.3879 [-0.6095; -0.1662]  0.01     0.53
Costa1997             acar   plac -0.8274 [-1.0401; -0.6147]  0.04     0.57
Hermansen2007         plac   sita  0.5700 [ 0.3170;  0.8230]  0.00     1.00
Garber2008            plac   vild  0.7000 [ 0.4505;  0.9495]  0.00     1.00
Alex1998              metf   sulf -0.6746 [-0.8482; -0.5011]  6.62     0.56
Johnston1994          migl   plac -0.9439 [-1.1927; -0.6952]  1.23     0.48
Johnston1998a         migl   plac -0.9439 [-1.1927; -0.6952]  4.35     0.32
Kim2007               metf   rosi  0.0877 [-0.0449;  0.2203]  0.14     0.08
Johnston1998b         migl   plac -0.9439 [-1.1927; -0.6952]  0.87     0.20
Gonzalez-Ortiz2004    metf   plac -1.1141 [-1.2309; -0.9973]  2.69     0.02
Stucci1996            benf   plac -0.9052 [-1.1543; -0.6561]  3.79     0.13
Moulin2006            benf   plac -0.9052 [-1.1543; -0.6561]  0.59     0.87
Willms1999            metf   plac -1.1141 [-1.2309; -0.9973]  0.04        .
Willms1999            acar   plac -0.8274 [-1.0401; -0.6147]  0.04        .

Number of studies: k = 26
Number of treatments: n = 10
Number of pairwise comparisons: m = 28
Number of designs: d = 15
```

```
Treatment estimate (sm = 'MD', comparison: other treatments vs 'plac'):
          MD              95%-CI      z  p-value
acar -0.8274 [-1.0401; -0.6147]  -7.62 < 0.0001
benf -0.9052 [-1.1543; -0.6561]  -7.12 < 0.0001
metf -1.1141 [-1.2309; -0.9973] -18.69 < 0.0001
migl -0.9439 [-1.1927; -0.6952]  -7.44 < 0.0001
piog -1.0664 [-1.2151; -0.9178] -14.06 < 0.0001
plac       .                  .       .        .
rosi -1.2018 [-1.2953; -1.1084] -25.22 < 0.0001
sita -0.5700 [-0.8230; -0.3170]  -4.42 < 0.0001
sulf -0.4395 [-0.6188; -0.2602]  -4.80 < 0.0001
vild -0.7000 [-0.9495; -0.4505]  -5.50 < 0.0001

Quantifying heterogeneity / inconsistency:
tau^2 = 0.1087; tau = 0.3297; I^2 = 81.4% [72.0%; 87.7%]

Tests of heterogeneity (within designs) and inconsistency (between designs):
                    Q d.f.  p-value
Total           96.99   18 < 0.0001
Within designs  74.46   11 < 0.0001
Between designs 22.53    7   0.0021
```

As we have created our network meta-analysis model, we can go ahead and draw our network graph (**Figure 2**).

Several types of information are conveyed by this network graph.

- First, there is the overall structure of comparisons in our network, which makes it possible for us to understand which treatments were compared with each other in the original data.

- Second, there are the edges having a different thickness, indicating how often this specific comparison can be found in our network. We see that there are many trials comparing Rosiglitazone with Placebo.

There is also one multiarm trial in our network, represented by the triangle shown in blue in our network.

As a next step, our attention can be shifted towards the direct and indirect evidence in our network by looking at the rate of direct and indirect contribution to each comparison. A function has been prepared to this end with the name of direct.evidence.plot.

As can be seen in **Figure 7**, there are many estimates included in our network model that needed to be inferred by indirect evidence only. We are also provided with two additional metrics by the plot: The Minimal Parallelism and the Mean Path

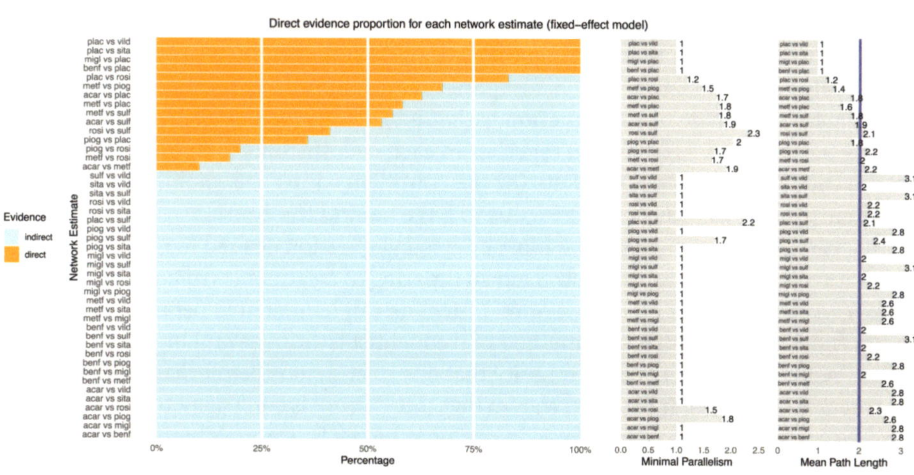

Figure 7.
Direct evidence proportion for each network estimate.

Length of each comparison. It is noted by König [49] that lower values of minimal parallelism and Mean Path Length >2 means that care should be taken while interpreting results for specific comparison.

Then we can look at our network's estimates for all possible combinations of treatments. In order to be able to do this, result matrices stored in our net meta results object under the fixed effects model can be used. Through a few preprocessing steps, the matrix can be made easier to read. First, the matrix is extracted from our data and the numbers in the matrix are rounded to three digits.

	acar	benf	metf	migl	piog	plac	rosi	sita	sulf	vild
acar	0.000	0.078	0.287	0.117	0.239	-0.827	0.374	-0.257	-0.388	-0.127
benf	-0.078	0.000	0.209	0.039	0.161	-0.905	0.297	-0.335	-0.466	-0.205
metf	-0.287	-0.209	0.000	-0.170	-0.048	-1.114	0.088	-0.544	-0.675	-0.414
migl	-0.117	-0.039	0.170	0.000	0.123	-0.944	0.258	-0.374	-0.504	-0.244
piog	-0.239	-0.161	0.048	-0.123	0.000	-1.066	0.135	-0.496	-0.627	-0.366
plac	0.827	0.905	1.114	0.944	1.066	0.000	1.202	0.570	0.439	0.700
rosi	-0.374	-0.297	-0.088	-0.258	-0.135	-1.202	0.000	-0.632	-0.762	-0.502
sita	0.257	0.335	0.544	0.374	0.496	-0.570	0.632	0.000	-0.131	0.130
sulf	0.388	0.466	0.675	0.504	0.627	-0.439	0.762	0.131	0.000	0.261
vild	0.127	0.205	0.414	0.244	0.366	-0.700	0.502	-0.130	-0.261	0.000

When the fact that a "triangle" in our matrix includes too much redundant information is considered, it seems to be possible to replace the lower triangle with an empty value.

	acar	benf	metf	migl	piog	plac	rosi	sita	sulf	vild
acar	"0"	"0.078"	"0.287"	"0.117"	"0.239"	"-0.827"	"0.374"	"-0.257"	"-0.388"	"-0.127"
benf	"."	"0"	"0.209"	"0.039"	"0.161"	"-0.905"	"0.297"	"-0.335"	"-0.466"	"-0.205"
metf	"."	"."	"0"	"-0.17"	"-0.048"	"-1.114"	"0.088"	"-0.544"	"-0.675"	"-0.414"
migl	"."	"."	"."	"0"	"0.123"	"-0.944"	"0.258"	"-0.374"	"-0.504"	"-0.244"
piog	"."	"."	"."	"."	"0"	"-1.066"	"0.135"	"-0.496"	"-0.627"	"-0.366"
plac	"."	"."	"."	"."	"."	"0"	"1.202"	"0.57"	"0.439"	"0.7"
rosi	"."	"."	"."	"."	"."	"."	"0"	"-0.632"	"-0.762"	"-0.502"
sita	"."	"."	"."	"."	"."	"."	"."	"0"	"-0.131"	"0.13"
sulf	"."	"."	"."	"."	"."	"."	"."	"."	"0"	"0.261"
vild	"."	"."	"."	"."	"."	"."	"."	"."	"."	"0"

The net league() function offers an extremely convenient way of exporting all estimated effect sizes. A matrix similar to the one given above can be generated by this function. Yet, in the matrix created by this function, only the pooled effect sizes belonging to the direct comparisons available in our network will be shown by the upper triangle, like the ones to be attained if a conventional meta-analysis had been conducted for each comparison. As there is no direct evidence for all comparisons, we will see some fields in the upper triangle empty. In this case, the network meta-analysis effect sizes for each comparison are contained by the lower triangle. The biggest advantage of this function is that it allows effect size estimates and confidence intervals to be shown together in each cell; the only thing that we need to tell the function is how the brackets for the confidence intervals should look like and how many digits we want our estimates to have behind the comma.

In a network meta-analysis, the most interesting question desired to be answered is: which intervention works the best? Such an ordering of treatments from most to least useful can be performed by the net rank() function implemented in net meta. The net rank() function is also built on a method of frequentist treatment ranking that uses P-scores. With these P-scores, the certainty that one treatment is better than another treatment is measured. It has been shown that this P-score is equivalent to the SUCRA score [50]. Our net meta object is needed as input by the function. Moreover, the small values parameter used to define whether smaller effect sizes in comparison are an indicator of a beneficial ("good") or harmful ("bad") effect should be specified. Now we will look at the output for our example:

```
      P-score
rosi  0.9789
metf  0.8513
piog  0.7686
migl  0.6200
benf  0.5727
acar  0.4792
vild  0.3512
sita  0.2386
sulf  0.1395
plac  0.0000
```

As can be seen, the Rosiglitazone treatment has the highest P-score, which indicates that this treatment may be particularly helpful. Contrarily, the P-score of Placebo is zero, supporting our intuition that placebo may not be the best treatment decision. It should be noted, however, that treatment should never be automatically concluded to be the best just because it has the highest score [51]. One of the good ways to be used to visualize the uncertainty in our network is to generate network forest plots with the "weakest" treatment as a comparison. The forest plot can also be used to do this. The reference group for the forest plot can be specified by using the reference group argument (**Figures 8** and **9**).

Now it can be seen that the results are more ambiguous than they seemed before; it is seen that several high-performing treatments having overlapping confidence intervals are available. This means we cannot make a firm judgment about which treatment is actually the best, but rather we see that there are a number of treatments that are more effective compared to placebo.

3.1 Decomposition of heterogeneity statistics

It is possible to decompose the Q total statistic (of the "whole network") into a Q statistic to assess heterogeneity between studies having the same design ("within designs") and a Q statistic to assess design inconsistency ("between designs"). The subsets of treatments that are compared with each other in a study are used to define designs.

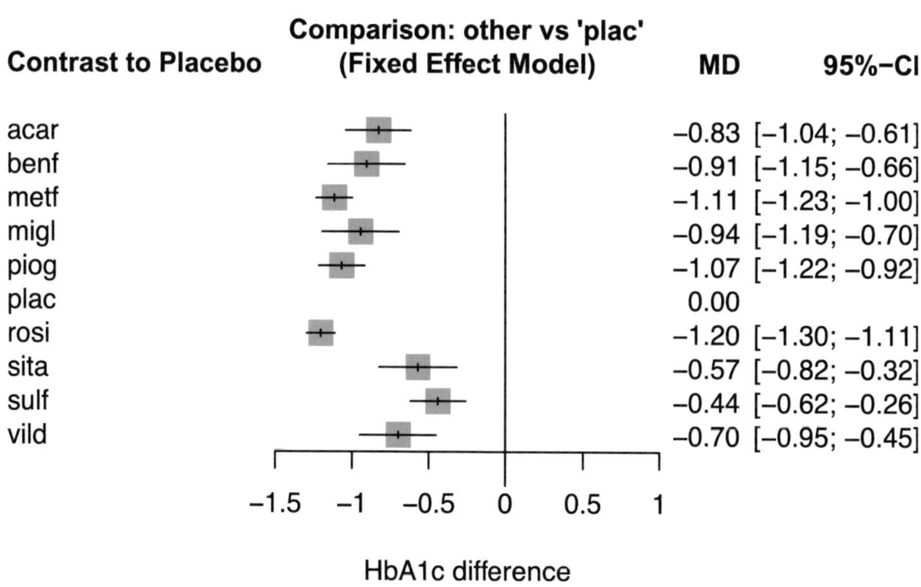

Figure 8.
Forest plot for fixed effect model with placebo as reference.

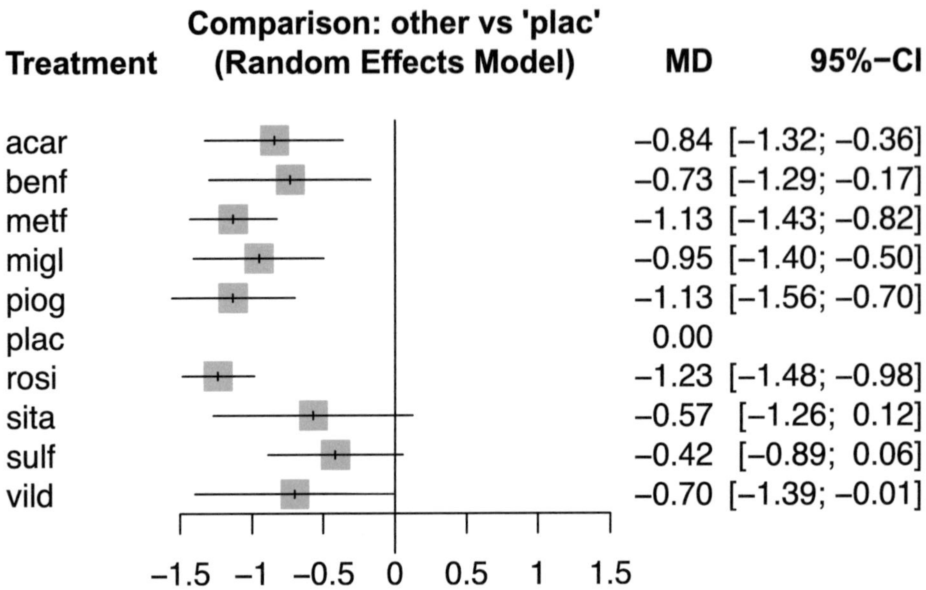

Figure 9.
Forest plot for random-effects model with placebo as reference.

```
                        Q df         pval
Total            96.98555 18 7.871369e-13
Within designs   74.45528 11 1.723944e-11
Between designs  22.53027  7 2.057014e-03
```

For this analysis, the fixed-effect model has been used and it is seen that there is considerable heterogeneity/inconsistency within as well as between designs. The total within-design heterogeneity can be further decomposed into the contribution from each design.

```
               design      Q df     pval
1        acar vs sulf  0.000  0       NA
2        metf vs piog  0.000  0       NA
3        metf vs rosi  0.187  1 6.65e-01
4        metf vs sulf  0.000  0       NA
5        piog vs rosi  0.000  0       NA
6        plac vs acar  0.000  0       NA
7        plac vs benf  4.381  1 3.63e-02
8        plac vs metf 42.164  2 6.98e-10
9        plac vs migl  6.449  2 3.98e-02
10       plac vs piog  0.000  0       NA
11       plac vs rosi 21.273  5 7.19e-04
12       plac vs sita  0.000  0       NA
13       plac vs vild  0.000  0       NA
14       rosi vs sulf  0.000  0       NA
15 plac vs acar vs metf  0.000  0       NA
```

As can be seen, the network meta-analysis includes 26 studies and these 26 studies use 15 different designs. Because only five designs for which more than one study exist, the remaining Q statistics specific to design are equal to zero and do not have any degrees of freedom. Except for design metf:rosi (p value = 0.67), heterogeneity is higher than would be expected between the contributing studies for all the other four designs; in the case of metf:plac a substantial amount more ($p < 0{:}0001$). Sources of this could be identified in a substantive application and thus the analysis could be updated appropriately.

```
                    Q df  pval tau.within
Between designs 2.194  7 0.948       0.38
```

Now the net heat plot, put forward by Krahn, König, and Binder [49] will be introduced. This is a graphical presentation where two types of information are shown in a single plot. These are:

1. For each network estimate, the contribution of each design to this estimate, and

2. For each network estimate, the extent of inconsistency due to each design.

Net heat plot is very useful in terms of evaluating the inconsistency in our network model, and what contributes to it (**Figure 10**).

A quadratic matrix is produced by the function so that each element in a row can be compared to all other elements in the columns. It should be noted here that rows and columns do not refer to all treatment comparisons in our network rather to specific designs. Thus, we also have rows and columns for the multiarm study, which had a design that compares "Plac", "Metf" and Acar. Comparison of treatments with only one type of evidence (i.e., indirect or indirect evidence) is not included in this chart, as we are dealing with cases of inconsistency between direct and indirect evidence. Moreover, the net heat plot has also two important properties: 1. Gray boxes. The Gray boxes for each design comparison show the extent to which one treatment comparison is important in terms of estimating another treatment comparison. The increasing size of the box indicates the increasing importance of comparison. This can be easily analyzed by going through the rows of the plot one after another, and then by checking for each row in which columns the gray boxes are the largest. In rows where the row comparison and the column comparison intersect, the boxes are large, which is a common finding and means that direct evidence was employed. For instance, it is possible to see a big gray box at the point where the "Plac vs Rosi2" row and the "Plac vc Rosi" column intersect [52].

The colored backgrounds which range from blue to red indicate the inconsistency of the comparison in a row, which can be attributed to the design in a column. Inconsistent fields are shown in the upper-left corner in red. For instance, it is seen that the entry in column "Metf vs. Sulf" is shown with red in the row for "Rosi vs. Sulf". This indicates that the evidence that "Metf vs. Sulf" provides for the "Metf vs. Sulf" estimation is not consistent with the other evidence. We can now remember

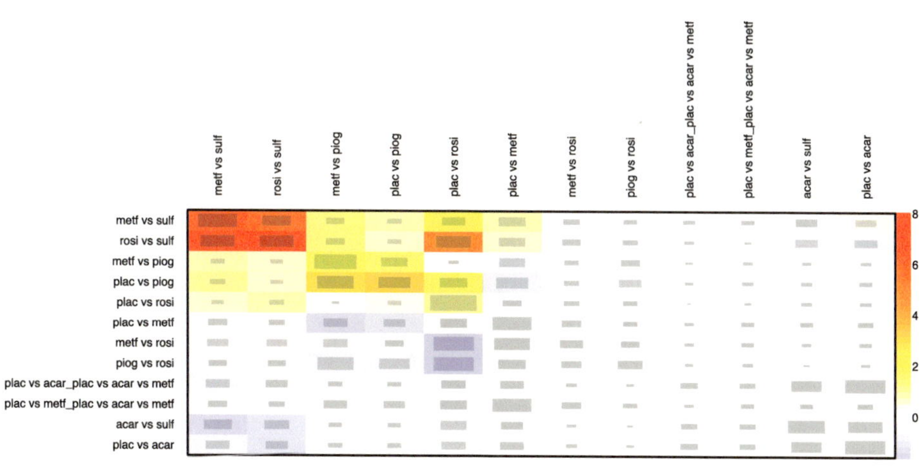

Figure 10.
Net heat plot of the Senn data example based on a fixed-effect model.

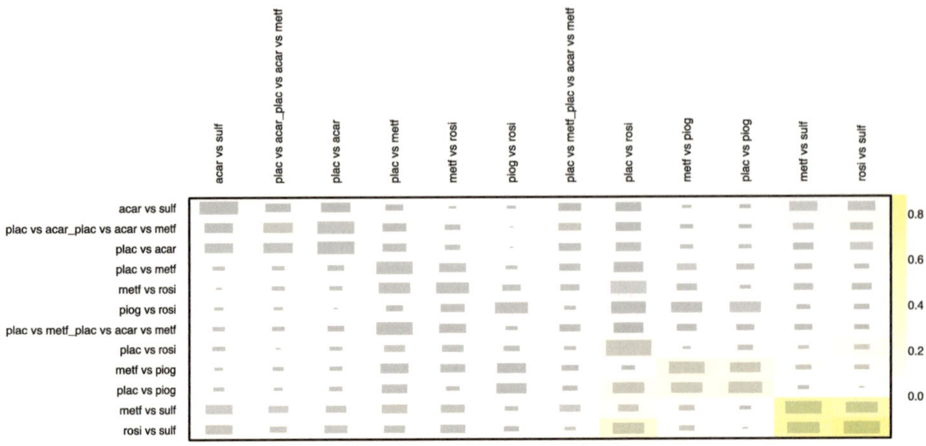

Figure 11.
Net heat plot of the Senn data example from a random-effects model.

that the fixed effects model that we initially used for our network analysis forms the basis of these results. On the basis of the things we have seen so far, we can reach the conclusion that due to too much unexpected heterogeneity, justification is not provided for the fixed effects model. How the net heat graph changes when a random-effects model is assumed can be controlled by changing the random argument of the net heat function to TRUE. It is seen that this results in a significant reduction of inconsistency in our network (**Figure 11**).

3.1.1 Net splitting

Net splitting, also known as node splitting, is another method for checking consistency in our network. With this method, our network estimates are split into the contribution of direct and indirect evidence and in this way, we can control for inconsistency in specific comparisons in our network. To generate a net split and compare the results.

```
Fixed effects model:

   comparison k prop     nma  direct   indir.    Diff     z  p-value
 acar vs benf 0    0  0.0778       .  0.0778       .      .        .
 acar vs metf 1 0.10  0.2867  0.2000  0.2966 -0.0966 -0.26   0.7981
 acar vs migl 0    0  0.1166       .  0.1166       .      .        .
 acar vs piog 0    0  0.2391       .  0.2391       .      .        .
 acar vs plac 2 0.63 -0.8274 -0.8172 -0.8446  0.0274  0.12   0.9030
 acar vs rosi 0    0  0.3745       .  0.3745       .      .        .
 acar vs sita 0    0 -0.2574       . -0.2574       .      .        .
 acar vs sulf 1 0.53 -0.3879 -0.4000 -0.3740 -0.0260 -0.11   0.9088
 acar vs vild 0    0 -0.1274       . -0.1274       .      .        .
 benf vs metf 0    0  0.2089       .  0.2089       .      .        .
 benf vs migl 0    0  0.0387       .  0.0387       .      .        .
 benf vs piog 0    0  0.1612       .  0.1612       .      .        .
 benf vs plac 2 1.00 -0.9052 -0.9052       .       .      .        .
 benf vs rosi 0    0  0.2967       .  0.2967       .      .        .
 benf vs sita 0    0 -0.3352       . -0.3352       .      .        .
 benf vs sulf 0    0 -0.4657       . -0.4657       .      .        .
 benf vs vild 0    0 -0.2052       . -0.2052       .      .        .
 metf vs migl 0    0 -0.1702       . -0.1702       .      .        .
 metf vs piog 1 0.68 -0.0477 -0.1600  0.1866 -0.3466 -2.32   0.0201
 metf vs plac 4 0.58 -1.1141 -1.1523 -1.0608 -0.0915 -0.76   0.4489
 metf vs rosi 2 0.18  0.0877  0.0731  0.0908 -0.0178 -0.10   0.9204
```

```
  metf vs sita 0    0 -0.5441         . -0.5441       .       .       .
  metf vs sulf 1 0.56 -0.6746 -0.3700 -1.0611  0.6911   3.88   0.0001
  metf vs vild 0    0 -0.4141         . -0.4141       .       .       .
  migl vs piog 0    0  0.1225         .  0.1225       .       .       .
  migl vs plac 3 1.00 -0.9439 -0.9439       .         .       .       .
  migl vs rosi 0    0  0.2579         .  0.2579       .       .       .
  migl vs sita 0    0 -0.3739         . -0.3739       .       .       .
  migl vs sulf 0    0 -0.5044         . -0.5044       .       .       .
  migl vs vild 0    0 -0.2439         . -0.2439       .       .       .
  piog vs plac 1 0.36 -1.0664 -1.3000 -0.9363 -0.3637 -2.30   0.0215
  piog vs rosi 1 0.20  0.1354  0.1000  0.1442 -0.0442 -0.22   0.8289
  piog vs sita 0    0 -0.4964         . -0.4964       .       .       .
  piog vs sulf 0    0 -0.6269         . -0.6269       .       .       .
  piog vs vild 0    0 -0.3664         . -0.3664       .       .       .
  rosi vs plac 6 0.83 -1.2018 -1.1483 -1.4665  0.3182   2.50   0.0125
  sita vs plac 1 1.00 -0.5700 -0.5700       .         .       .       .
  sulf vs plac 0    0 -0.4395         . -0.4395       .       .       .
  vild vs plac 1 1.00 -0.7000 -0.7000       .         .       .       .
  rosi vs sita 0    0 -0.6318         . -0.6318       .       .       .
  rosi vs sulf 1 0.41 -0.7623 -1.2000 -0.4575 -0.7425 -3.97 < 0.0001
  rosi vs vild 0    0 -0.5018         . -0.5018       .       .       .
  sita vs sulf 0    0 -0.1305         . -0.1305       .       .       .
  sita vs vild 0    0  0.1300         .  0.1300       .       .       .
  sulf vs vild 0    0  0.2605         .  0.2605       .       .       .

Legend:
 comparison - Treatment comparison
 k          - Number of studies providing direct evidence
 prop       - Direct evidence proportion
 nma        - Estimated treatment effect (MD) in network meta-analysis
 direct     - Estimated treatment effect (MD) derived from direct evidence
 indir.     - Estimated treatment effect (MD) derived from indirect evidence
 Diff       - Difference between direct and indirect treatment estimates
 z          - z-value of test for disagreement (direct versus indirect)
```

Here, the important information is found in the p-value column. Any value that is $p < 0.05$ in this column is an indicator of a significant discrepancy (inconsistency) between the direct and indirect estimates. In the output, it is seen that there are indeed few comparisons showing significant discrepancies between direct and indirect evidence when the fixed effects model is used. Net split results can be visualized with a forest chart showing all comparisons for which both direct and indirect evidence are present in **Figure 12**.

4. Conclusions

For the estimation and comparison of treatment effects in a particular area, network meta-analysis can be used as a potentially powerful tool for using all the evidence. This approach has been depicted through an example from diabetes [48], which shows how to graph the network and explore a range of analyses. The results of our first model (fixed-effect model) Q value of DeFronzo1995 is highest with $Q = 30.89$. As a network model, the effects of all treatments are displayed in comparison to the placebo condition, which is why there is no effect shown for placebo. We can say heterogeneity/inconsistency in our network model is high, with $I^2 = 84\%$. The heterogeneity between treatment designs reflects the actual inconsistency in our network, and is highly significant ($p = 0.0021$). In **Figure 2**, looking at the network graph, it is seen that Rosiglitazone has been compared to Placebo in many trials. The only multi-arm trial in our network is that of Willms 2003. We see that it is the Rosiglitazone treatment with the highest P score. It is necessary to look at network forest plots with the "weakest" treatment, as it can be misleading to conclude that a treatment is best just because it has the highest score.

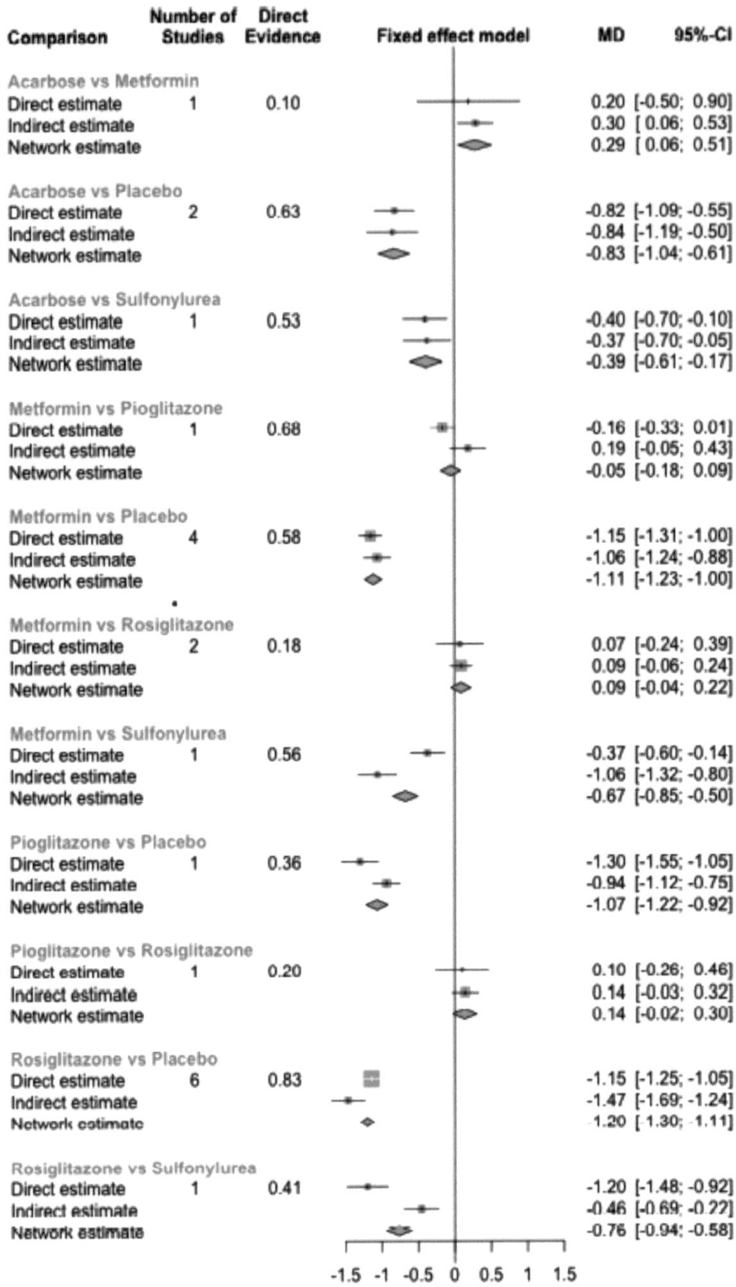

Figure 12.
Net split plot of the Senn data example from a fixed-effect model.

Looking at the forest network plot, we see that there are several high-performance treatments with overlapping confidence intervals. From here, we looked at the net heat plot as we could not make a definitive decision.

The extent of the information obtained in a given treatment comparison by means of indirect evidence and the extent of heterogeneity can be defined as two important aspects of network meta-analysis. The net heat graph communicates information about both of these and the software allows for the decomposition of heterogeneity within and between designs. If there is clinically relevant heterogeneity, it is worth being explored further. Looking at **Figure 10**, a particularly large gray box is seen where the "Plac vs. Rosi2 row and the "Plac vs. Rosi" column intersect. Using the.random-effects model in **Figure 11**, we see that the inconsistency is significantly reduced.

Since it is not possible to conduct covariate adjustment at present with the software, one approach is to conduct study-specific (ideally individual participant data) analyses with appropriate covariate adjustment before the software presented here is used to perform network meta-analysis.

Author details

Nilgün Yildiz
Marmara University, Istanbul, Turkey

*Address all correspondence to: ncelebi@marmara.edu.tr

References

[1] Shim SR, Yoon BY, Shin IS, Bae JM. Network meta-analysis: Application and practice using Stata. Korean Society of Epidemiology. 2017;**39**:e2017047. DOI: 10.4178/epih.e2017047

[2] Caldwell DM, Ades AE, Higgins JP. Simultaneous comparison of multiple treatments: Combining direct and indirect evidence. BMJ. 2005;**331**:897-900

[3] Li T, Vedula SS, Scherer R, Dickersin K. What comparative effectiveness research is needed? A framework for using guidelines and systematic reviews to identify evidence gaps and research priorities. Annals of Internal Medicine. 2012;**156**:367-377

[4] Mitka M. US government kicks off program for comparative effectiveness research. Journal of the American Medical Association. 2010;**304**:2230-2231

[5] Lu G, Ades AE. Assessing evidence inconsistency in mixed treatment comparisons. Journal of the American Statistical Association. 2006;**101**: 447-459

[6] Salanti G, Higgins JP, Ades AE, Ioannidis JP. Evaluation of networks of randomized trials. Statistical Methods in Medical Research. 2008;**17**:279-301

[7] Higgins JP, Whitehead A. Borrowing strength from external trials in a meta-analysis. Statistics in Medicine. 1996;**15**: 2733-2749

[8] Mills EJ, Ioannidis JP, Thorlund K, Schünemann HJ, Puhan MA, Guyatt GH. How to use an article reporting a multiple treatment comparison metaanalysis. Journal of the American Medical Association. 2012; **308**:1246-1253

[9] Bucher HC, Guyatt GH, Griffith LE, Walter SD. The results of direct and indirect treatment comparisons in meta-analysis of randomized controlled trials. Journal of Clinical Epidemiology. 1997;**50**:683-691

[10] Mills EJ, Ghement I, O'Regan C, Thorlund K. Estimating the power of indirect comparisons: A simulation study. PLoS One. 2011;**6**(1):e16237

[11] Ioannidis JP. Indirect comparisons: The mesh and mess of clinical trials. Lancet. 2006;**368**:1470-1472

[12] Cipriani A, Higgins JP, Geddes JR, Salanti G. Conceptual and technical challenges in network meta-analysis. Annals of Internal Medicine. 2013;**159**: 130-137

[13] Tonin FS, Rotta I, Mendes AM, Pontarolo R. Network meta-analysis: A technique to gather evidence from direct and indirect comparisons. Pharmacy Practice (Granada). 2017;**15**:943

[14] Hoaglin DC, Hawkins N, Jansen JP, Scott DA, Itzler R, Cappelleri JC, et al. Conducting indirect-treatment-comparison and network meta-analysis studies: Report of the ISPOR Task Force on indirect treatment comparisons good research practices: Part 2. Value in Health. 2011;**14**:429-437

[15] Li T, Puhan MA, Vedula SS, Singh S, Dickersin K, Ad Hoc Network Meta-analysis Methods Meeting Working Group. Network meta-analysis-highly attractive but more methodological research is needed. BMC Medicine. 2011;**9**:79

[16] Mills EJ, Bansback N, Ghement I, Thorlund K, Kelly S, Puhan MA, et al. Multiple treatment comparison meta-analyses: A step forward into complexity. Clinical Epidemiology. 2011;**3**:193-202

[17] Reken S, Sturtz S, Kiefer C, Böhler YB, Wieseler B. Assumptions of

mixed treatment comparisons in health technology assessments: Challenges and possible steps for practical application. PLoS One. 2016;**11**:e0160712

[18] Veroniki AA, Vasiliadis HS, Higgins JP, Salanti G. Evaluation of inconsistency in networks of interventions. International Journal of Epidemiology. 2013;**42**:332-345

[19] Bhatnagar N, Lakshmi PV, Jeyashree K. Multiple treatment and indirect treatment comparisons: An overview of network meta-analysis. Perspectives in Clinical Research. 2014; **5**:154-158

[20] Mills EJ, Thorlund K, Ioannidis JP. Demystifying trial networks and network meta-analysis. BMJ. 2013;**346**:f2914

[21] Salanti G. Indirect and mixed-treatment comparison, network, or multiple-treatments meta-analysis: Many names, many benefits, many concerns for the next generation evidence synthesis tool. Research Synthesis Methods. 2012;**3**:80-97

[22] Lu G, Ades AE. Combination of direct and indirect evidence in mixed treatment comparisons. Statistics in Medicine. 2004;**23**:3105-3124

[23] Jansen JP, Fleurence R, Devine B, Itzler R, Barrett A, Hawkins N, et al. Interpreting indirect treatment comparisons and network meta-analysis for health-care decision making: Report of the ISPOR Task Force on indirect treatment comparisons good research practices: Part 1. Value in Health. 2011; **14**:417-428

[24] Jansen JP, Naci H. Is network meta-analysis as valid as standard pairwise meta-analysis? It all depends on the distribution of effect modifiers. BMC Medicine. 2013;**11**:159

[25] Dakin HA, Welton NJ, Ades AE, Collins S, Orme M, Kelly S. Mixed treatment comparison of repeated measurements of a continuous endpoint: An example using topical treatments for primary openangle glaucoma and ocular hypertension. Statistics in Medicine. 2011;**30**:2511-2535

[26] Schmitz S, Adams R, Walsh CD, Barry M, FitzGerald O. A mixed treatment comparison of the efficacy of anti-TNF agents in rheumatoid arthritis for methotrexate non-responders demonstrates differences between treatments: A Bayesian approach. Annals of the Rheumatic Diseases. 2012; **71**:225-230

[27] Jones B, Roger J, Lane PW, Lawton A, Fletcher C, Cappelleri JC, et al. Statistical approaches for conducting network meta-analysis in drug development. Pharmaceutical Statistics. 2011;**10**:523-531

[28] White IR. Network meta-analysis. The Stata Journal. 2015;**15**:951-985

[29] Cooper NJ, Peters J, Lai MC, et al. How valuable are multiple treatment comparison methods in evidence-based health-care evaluation? Value in Health. 2011;**14**(2):371-380

[30] Edwards SJ, Clarke MJ, Wordsworth S, Borrill J. Indirect comparisons of treatments based on systematic reviews of randomised controlled trials. International Journal of Clinical Practice. 2009;**63**:841-854. DOI: 10.1111/ j.1742-1241.2009.02072

[31] Gartlehner G, Moore CG. Direct versus indirect comparisons: A summary of the evidence. The International Journal of Technology Assessment in Health Care. 2008;**24**: 170-177. DOI: 10.1017/S02664623080 80240

[32] Efthimiou O, Debray TPA, vanValkenhoef G, Trelle S, Panayidou K, Moons KGM, et al. GetReal in network meta-analysis: A

review of the methodology. Research Synthesis Methods. 2016;7:236-263. DOI: 10.1002/jrsm.1195

[33] Salanti G, Kavvoura FK, Ioannidis JP. Exploring the geometry of treatment networks. Annals of Internal Medicine. 2008;**148**:544-553

[34] Higgins JPT, Jackson D, Barrett JK, Lu G, Ades AE, White IR. Consistency and inconsistency in network meta-analysis: Concepts and models for multi-arm studies. Research Synthesis Methods. 2012;**3**:98-110

[35] Higgins JPT. Commentary: Heterogeneity in meta-analysis should be expected and appropriately quantified. International Journal of Epidemiology. 2008;**37**:1158-1160

[36] Chan AW, Altman DG. Epidemiology and reporting of randomised trials published in PubMed journals. The Lancet. 2005;**365**:1159-1162

[37] Dias S, Welton NJ, Caldwell DM, Ades AE. Checking consistency in mixed treatment comparison meta-analysis. Statistics in Medicine. 2010b;**29**:932-944. DOI: 10.1002/sim.3767

[38] Song F, Loke YK, Walsh T, Glenny AM, Eastwood AJ, Altman DG. Methodological problems in the use of indirect comparisons for evaluating healthcare interventions: Survey of published systematic reviews. BMJ. 2009;**338**:b1147

[39] Baker SG, Kramer BS. The transitive fallacy for randomized trials: If A bests B and B bests C in separate trials, is A better than C? BMC Medical Research Methodology. 2002;**2**:13

[40] Donegan S, Williamson P, Gamble C, Tudur SC. Indirect comparisons: A review of reporting and methodological quality. PLoS One. 2010;**5**:e11054. DOI: 10.1371

[41] Song F, Altman DG, Glenny AM, Deeks JJ. Validity of indirect comparison for estimating efficacy of competing interventions: empirical evidence from published meta-analyses. BMJ. 2003;**326**:472. DOI: 10.1136/bmj.326.7387.472

[42] Dias S, Welton NJ, Sutton AJ, Caldwell DM, Lu G, Ades AE. Evidence synthesis for decision making 4: Inconsistency in networks of evidence based on randomized controlled trials. Medical Decision Making. 2013d;**33**:641-656. DOI: 10.1177/0272989X12455847

[43] Donegan S, Williamson P, D'Alessandro U, Tudur SC. Assessing key assumptions of network meta-analysis: A review of methods. Research Synthesis Methods. 2013b;**4**:291-323. DOI: 10.1002/jrsm.1085

[44] Salanti G, Marinho V, Higgins JP. A case study of multiple-treatments meta-analysis demonstrates that covariates should be considered. Journal of Clinical Epidemiology. 2009;**62**:857-864. DOI: 10.1016/j.jclinepi.2008.10.001

[45] Julious SA, Wang SJ. How biased are indirect comparisons, particularly when comparisons are made over time in controlled trials. Drug Information Journal. 2008;**42**:625

[46] Lu G, Ades A. Modeling between-trial variance structure in mixed treatment comparisons. Biostatistics. 2009;**10**:792-805. DOI: 10.1093/biostatistics/kxp032

[47] Lumley T. Network meta-analysis for indirect treatment comparisons. Statistics in Medicine. 2002;**21**:2313-2324. DOI: 10.1002/sim.1201

[48] Senn S, Gavini F, Magrez D, Scheen A. Issues in performing a network meta-analysis. Statistical Methods in Medical Research. 2013;**22**(2):169-189. DOI: 10.1177/0962280211432220

[49] König J, Krahn U, Binder H. Visualizing the flow of evidence in network meta-analysis and characterizing mixed treatment comparisons. Statistics in Medicine. 2013;**32**(30):5414-5429

[50] Rücker G, Schwarzer G. Ranking treatments in frequentist network meta-analysis works without resampling methods. BMC Medical Research Methodology. 2015;**15**(1):58

[51] Mbuagbaw L, Rochwerg B, Jaeschke R, Heels-Andsell D, Alhazzani W, Thabane L, et al. Approaches to interpreting and choosing the best treatments in network meta-analyses. Systematic Reviews. 2017;**6**(1):79

[52] Schwarzer G, Carpenter JR, Rücker G. Meta-Analysis with R. Switzerland: Springer International Publishing; 2015

Chapter 8

Variance Balanced Design

D.K. Ghosh

Abstract

In this chapter binary, ternary and n-ary variance balanced design is constructed using balanced incomplete block, resolvable balanced incomplete block, semi regular group divisible, factorial, fractional factorial designs. Constructed variance balanced designs are with v, (v + 1), (v + 2) and (v + r) treatments. Method of construction of variance balanced designs are supported by suitable examples. It is found that all most all variance balanced designs are with high efficiency factors.

Keywords: incidence matrix, C – Matrix, resolvable balanced incomplete block designs, eigen values, balanced and group divisible designs

1. Introduction

In literature balanced incomplete block designs are either variance balanced (VB), efficiency balanced (EB) or pairwise balanced. Raghvarao ([1], Theorem 4.5.2) discussed that among the class of connected designs the balanced designs are the most efficient designs. A design is said to be variance balanced, if the variance of the estimate of each of the possible elementary treatment contrast is the same, i.e., if t_i denotes the estimate of ith treatment effects, then Var $(t_i - t_j)$ is constant for all $i \neq j$.

Chakrabarti [2] gave useful concept of C – matrix of design. It is known that balanced incomplete block designs are the most efficient but do not exist for all parametric specifications, and they are equi replicated and have equal block sizes. In some situations, balanced block designs with equal replicates or unequal block size or both are needed. The variance balanced designs can have both equal and unequal number of replications and block sizes. The importance of variance balanced designs in the context of experimental material is well known, as it yields optimal designs apart from ensuring simplicity in the analysis. Many practical situations demand designs with varying block sizes (Pearce, [3], or resolvable VB designs with unequal replications Mukerjee and Kageyama [4]). Rao [5], Headyat and Federer [6], Raghavarao [7] and Puri and Nigam [8] defined that a design is said to be variance balanced, if every normalized estimable linear function of treatment effect can be estimated with the same precision. They also discussed the necessary and sufficient conditions for the existence of such designs. John [9], Jones et al. [10], Kageyama [11, 12], Kageyama et al. [13], Pal and Pal [14], Roy [15], Sinha [16, 17], Sinha and Jones [18] and Tyagi [19] gave some more methods for constructing block designs with unequal treatment replications and unequal block sizes. Khatri [20], along with a method of construction of VB designs, gave a formula to measure over-all A-efficiency of variance balanced designs. Das and Ghosh [21] gave the methods of construction of variance balanced designs with augmented blocks and treatments. Mukerjee and Kageyama [22] introduced resolvable variance balanced designs. A technique for constructing variance balanced designs, which is based on the unionizing block principle of Headayat and Federer [6], was described in Calvin [23]. Calvin and Sinha [24] extended his technique to produce designs with more than two distinct block sizes that permit fewer replications. Agarwal and Kumar [25] gave a method of construction of variance balanced designs which is associated with group divisible (GD) designs. Rao [5] observed that, if the information matrix C of a block design satisfied

$$C = \theta\left[I_v - \frac{1}{v}E_{vv}\right]$$

where, θ is non zero eigen value of C matrix, I_v is an identity matrix of order v, E_{vv} is the matrix with v rows and v columns where, all the elements are unity, then such design is called Variance balanced designs. Since balanced incomplete block design (BIBD) satisfies this property and hence, balanced incomplete block design is a particular case of Variance balanced designs.

Das and Ghosh [21] defined generalized efficiency balanced (GEB) design, which include both VB as well as EB designs. Ghosh [26], Ghosh and Karmaker [27], Ghosh and Devecha [28], Ghosh, Divecha, and Kageyama [29], Ghosh et al. [30], Ghosh, et al. [31, 32] obtained several methods for construction of VB designs. Ghosh and Joshi [30] constructed VB design through GD design. Again, Ghosh and Joshi [33] Constructed VB Design through Triangular design. Kageyama [10] recommended the use of non-binary VB design, when binary VB designs are not available for given values of parameters. Ghosh and Ahuja [34] carried out VB design using fractional factorial designs. Agarwal and Kumar [35, 36] developed some methods of constructing ternary VB designs with $(v + s)$ treatments $(s \geq 1)$, having blocks of unequal sizes, through block designs with v treatment. Ghosh, Kageyama and Joshi [37] developed Ternary VB designs using BIB and GD design. Ghosh et al. [37] further obtained more VB designs using Latin square type PBIB design. Ghosh [38] studied the robustness of variance balanced design against the loss of k treatments and one block. Ghosh et al. [39] discuss construction of VB design using factorial designs. Hedayat and Stufken [40] established a relation between pair wise balanced and variance balanced designs. Jones [41] discussed the property of incomplete block designs. Gupta and Jones [42] constructed equal replicated VB designs.

2. Method of construction

Method of construction of Variance balanced design with equal/unequal replication sizes and equal/unequal block sizes is carried pot in this chapter. Section 3 discusses the construction of variance balanced design using Hadamard matrix. While construction of variance balanced design using semi regular group divisible design is discussed in Section 4. Variance balanced design is constructed by augmenting n more blocks which is discussed in Section 5. Construction of variance balanced design with (v + 1) treatments using unreduced balanced incomplete block design is shown in Section 6. Section 7 discusses the construction of variance balanced design using 2^n symmetrical factorial experiments. Variance balanced design is constructed using incidence matrix also, and is shown in section – 8.

3. Variance balanced design using Hadamard matrix

Theorem – 3.1: Equi-replicated Variance balanced design with parameters v = n – 1, b = n, r = n/2, k = {n – 1, n/2–1} and $C_{im} = \frac{(3n-4)}{(n-1)(n-2)}$ can always be constructed from a Hadamard matrix of size n by deleting its first row and then considering rows as treatments and columns as blocks.

Proof: Consider a Hadamard matrix of size n. Delete its first row. The size of this matrix become (n – 1) x n. We replace −1 by 0, and call this matrix by N. This matrix contains (n-1) rows and n columns where, element "1" occurs n/2 times in each row, (n – 1) times in first column, and (n/2–1) times in the remaining columns. Consider matrix N as an incidence matrix of a variance balanced design, where rows are treatments and columns are blocks, so, v = n-1, b = n, r = n/2 and k = {n - 1, n/2–1} .

For variance balanced design, $C_{im} = \sum_j^b \frac{n_{ij}n_{mj}}{n.j}$, where, $i \neq m$ = 1 to v.

C_{im} is computed as $C_{1m} = \frac{1}{n-1} + \frac{1}{\frac{n}{2}-1} = \frac{(3n-4)}{(n-1)(n-2)}$.

We can verify that, C_{im} gives same constant value for each pair of treatments. Now a block design Is said to variance balanced design, if C matrix satisfies, C = θ (I_v – E_{vv}/v), where, θ is non zero eigen value of C matrix with multiplicity (v – 1), where,

$$C = \text{diag}(r_1, r_2, \ldots, r_v) – N\,K^{-1}N'.$$

$$C = \begin{bmatrix} n/2 & 0 & .. & 0 \\ 0 & \frac{n}{2} & .. & 0 \\ : & : & : & : \\ 0 & 0 & .. & n/2 \end{bmatrix} - \begin{bmatrix} n(n-2) & 3n-4 & .. & 3n-4 \\ 3n-4 & n(n-2) & .. & 3n-4 \\ : & : & .. & : \\ 3n-4 & 3n-4 & .. & n(n-2) \end{bmatrix} /(n–1)(n–2)$$

After simplification C reduces to

$$C = \frac{n(n-2)(n-3) + 6n - 8}{2(n-1)(n-2)}\left[I_v - \frac{E_{vv}}{v}\right] \tag{1}$$

Where, $\theta = \frac{n(n-2)(n-3)+6n-8}{2(n-1)(n-2)}$ denotes the non-zero eigen value of C matrix with multiplicity (n – 2).

Eq. (1) satisfy the condition of variance balanced design. Hence, this is an equi-replicated and two unequal block sizes variance balanced design.

3.1 Efficiency factor of a variance balanced design

The efficiency factor of a variance balanced design is defined as

$$E = \frac{\text{Var}(\hat{t_i} - \hat{t_m})_{RBD}}{\text{Var}(\hat{t_i} - \hat{t_m})_{VB}}$$

Where, $\text{Var}(\hat{t_i} - \hat{t_m})_{RBD}$ = (2/r) $\sigma^2 = \frac{2}{n/2}\sigma^2$ and

$$\text{Var}(\hat{t_i} - \hat{t_m})_{VB} = (2/\theta)\sigma^2 = \frac{2}{\frac{n(n-2)(n-3)+6n-8}{2(n-1)(n-2)}}\sigma^2$$

$$E = \frac{n(n-2)(n-3) + 6n - 8}{n(n-1)((n-2)}$$

Example–3.1 Construct a variance balanced design from a Hadamard matrix of size 8.

Using Theorem – 3.1, we construct a variance balanced design from a Hadamard matrix of size 8 as following:

Hadamard Matrix of size 8	Incidence matrix of a Variance balanced design
1 1 1 1 1 1 1 1.	
1 -1 1 -1 1 -1 1 -1	1 0 1 0 1 0 1 0
1 1 -1 -1 1 1 -1 -1	1 1 0 0 1 1 0 0
1 -1 -1 1 1 -1 -1 1 N = 1	0 0 1 1 0 0 1
1 1 1 1 -1 -1 -1 -1	1 1 1 1 0 0 0 0
1 -1 1 -1 -1 1 -1 1	1 0 1 0 0 1 0 1
1 1 -1 -1 -1 -1 1 1	1 1 0 0 0 0 1 1
1 -1 -1 1 -1 1 1 -1	1 0 0 1 0 1 1 0

N gives the incidence matrix of an equi-replicated and un equal block sizes variance balanced design with parameters v = 7, b = 8, r = 4, k = {7, 3}, C_{im} = 10/21 and information matrix,

$$C = \begin{bmatrix} 4 & 0 & .. & 0 \\ 0 & 4 & .. & 0 \\ \vdots & \vdots & \vdots & \vdots \\ 0 & 0 & .. & 4 \end{bmatrix} - \begin{bmatrix} 48 & 20 & .. & 20 \\ 20 & 48 & .. & 20 \\ \vdots & \vdots & .. & \vdots \\ 20 & 20 & .. & 48 \end{bmatrix} /42$$

After simplification, C reduces to

$$C = \frac{280}{84}\left[I_7 - \frac{E_{77}}{7}\right] \quad = \theta\left[I_7 - \frac{E_{77}}{7}\right] \tag{2}$$

Where, $\theta = \frac{10}{3}$, is the non zero eigen value of C matrix with multiplicity 6. Hence, it is a variance balanced design, with $\widehat{t_i} = (1/\theta)Q_i = (3/10)Q_i$,

$$\text{Var}\left(\widehat{t_i} - \widehat{t_m}\right)_{VB} = (2/\theta)\sigma^2 = (6/10)\,\sigma^2.\text{Var}\left(\widehat{t_i} - \widehat{t_m}\right)_{RBD} = (2/r)\,\sigma^2 = \frac{2}{4}\,\sigma^2$$

and Efficiency factor, E = 5/6. This shows that efficiency factor is very high.

4. Variance balanced design through semi regular group divisible designs

In this section, we discuss the construction of variance balanced design by adding the blocks of semi-regular group divisible design with its groups, provided the following conditions (i) block sizes, $k = \lambda_2$, (ii) $\lambda_1 = 0$ and (iii) number of groups are considered as number of blocks, are satisfied.

Theorem – 4.1 Let the parameters of a semi regular group divisible design are v, b, r, k, $\lambda_1 = 0$, λ_2, m and n, where $k = \lambda_2$. By adding the b blocks of this semi regular group divisible design with number of groups as blocks, an equi-replicated and un-equal block sizes variance balanced design is constructed with parameters $v_1 = v$, $b_1 = b + mn$, $r_1 = r + n$, $k_1 = \{k, n\}$ and $C_{im} = \lambda_2 / k$ or $C_{im} = \lambda_1 / k + n/n$.

Proof: Consider a semi regular group divisible design with parameters v, b, r, k, $\lambda_1 = 0$, $\lambda_2 = k$, m and n, where, m denotes number of groups and n number of treatments per group. Denote N as the incidence matrix of the resulting design. Consider one group as one block. Here, there are m groups and hence, we have m more blocks. Add b blocks of the semi-regular group divisible design with its m more blocks, provided m blocks are repeated n times. Hence, $v_1 = v$, $b_1 = b + mn$,

$r_1 = r + n$ and $k_1 = \{k, n\}$. We can check, $C_{im} = \sum_j^b \frac{n_{ij}n_{mj}}{n_{.j}}$, where $i \neq m = 1$ to v, for each pair of treatment as following. $C_{1m} = \frac{\lambda_2}{k}$, for those pair of treatments, which occur λ_2 times. Since, $\lambda_2 = k$ and hence, $C_{im} = 1$. Again, for those pair of treatments for which $\lambda_1 = 0$, $C_{1m} = \frac{\lambda_1}{k} + \frac{n}{n} = 1$. For variance balanced design, C_{im} should be the same for each pair of treatments. Hence, $\frac{\lambda_1}{k} + \frac{n}{n} = \frac{\lambda_2}{k}$. This implies that, using this method, we can construct a variance balanced design from those semi - regular group divisible designs in which $(\lambda_2 - \lambda_1) = k$ holds true.

Again, a block design Is said to variance balanced design, if C matrix satisfies, $C = \theta\,(I_v - E_{vv}/v)$, where, θ is non zero eigen value of C matrix with multiplicity $(v - 1)$, and, $C = \text{diag}\,(r_1, r_2, \ldots, r_v) - N\,K^{-1}N'$.

$$C = \begin{bmatrix} r+n & 0 & .. & 0 \\ 0 & r+n & .. & 0 \\ : & : & : & : \\ 0 & 0 & .. & r+n \end{bmatrix} - \begin{bmatrix} \frac{r+k}{k} & 1 & .. & 1 \\ 1 & \frac{r+k}{k} & .. & 1 \\ : & : & .. & : \\ 1 & 1 & .. & \frac{r+k}{k} \end{bmatrix}$$

Diagonal elements = $[k\,(r + n) - (r + k)]/k$, and off diagonal elements = $- k/k = -1$. After simplification, C reduces to

$$C = \frac{k(r+n) - r}{k}\left[I_v - \frac{E_{vv}}{v}\right] \tag{3}$$

Where, $\theta = \frac{k(r+n)-r}{k}$ denotes the non-zero eigen value of C matrix with multiplicity $(v - 1)$.

Eq. (3) satisfy the condition of variance balanced design. Hence, this is equi replicated and two unequal block sizes variance balanced design with parameters $v_1 = v$, $b_1 = b + mn$, $r_1 = r + n$, $k_1 = \{k, n\}$.

4.1 Efficiency factor

The efficiency factor of a variance balanced design is defined as

$$E = \frac{\text{Var}(\hat{t_i} - \hat{t_m})_{RBD}}{\text{Var}(\hat{t_i} - \hat{t_m})_{VB}}, \text{ where, } \text{Var}(\hat{t_i} - \hat{t_m})_{RBD} = (2/r)\,\sigma^2 = \frac{2}{r+n}\,\sigma^2,$$

$$\text{Var}(\hat{t_i} - \hat{t_m})_{VB} = (2/\theta)\sigma^2 = \frac{2}{\frac{k(r+n)-r}{k}}\sigma^2 \text{ and } E = \frac{k(r+n)\,) - r}{k(r+n)}$$

Example – 4.1 Construct a variance balanced design with parameters $v_1 = 6$, $b_1 = 18$, $r_1 = 8$, $k_1 = \{3, 2\}$ from a semi regular group divisible design SR – 20, having parameters $v = 6$, $b = 12$, $r = 6$, $k = 3$, $\lambda_1 = 0$, $\lambda_2 = 3$, $m = 3$ and $n = 2$. Where, group is (3,2).

Three groups each with 2 treatments are (1 4), (2 5), (3 6).

Blocks of the semi-regular group divisible design, SR – 20 are.

(1 2 3), (2 4 6), (3 4 5), (1 5 6), (1 2 3), (2 4 6), (3 4 5), (1 5 6), (1 2 6), (1 3 5), (2 3 4), (4 5 6),

Using Theorem – 4.1, incidence matrix of the variance balanced design is given as.

$$N_1 = \begin{matrix} 1&0&0&1&1&0&0&1&1&1&0&0&1&0&0&1&0&0 \\ 1&1&0&0&1&1&0&0&1&0&1&0&0&1&0&0&1&0 \\ 1&0&1&0&1&0&1&0&0&1&1&0&0&0&1&0&0&1 \\ 0&1&1&0&0&1&1&0&0&0&1&1&1&0&0&1&0&0 \\ 0&0&1&1&0&0&1&1&0&1&0&1&0&1&0&0&1&0 \\ 0&1&0&1&0&1&0&1&1&0&0&1&0&0&1&0&0&1 \end{matrix}$$

N_1 gives the incidence matrix of an equi-replicated and un-equal block sizes variance balanced design with parameters $v_1 = 6$, $b_1 = 18$, $r_1 = 8$, $k_1 = \{3, 2\}$ $C_{im} = 1$ and information matrix,

$$C = \begin{bmatrix} 8 & 0 & .. & 0 \\ 0 & 8 & .. & 0 \\ : & : & : & : \\ 0 & 0 & .. & 8 \end{bmatrix} - \begin{bmatrix} 18 & 6 & .. & 6 \\ 6 & 18 & .. & 6 \\ : & : & .. & : \\ 6 & 6 & .. & 18 \end{bmatrix} /6$$

After simplification, C reduces to

$$C = 6\left[I_7 - \frac{E_{77}}{7}\right] \quad = \theta\left[I_6 - \frac{E_6}{6}\right]$$

Where, $\theta = 6$ is the non zero eigen value of C matrix with multiplicity 5. Hence, it is a variance balance design with $\hat{t_i} = (1/\theta)\, Q_i = (1/6)\, Q_i$, $\mathrm{Var}(\hat{t_i} - \hat{t_m})_{VB} = (2/\theta)\sigma^2 = (2/6)\sigma^2$, $\mathrm{Var}(\hat{t_i} - \hat{t_m})_{RBD} = (2/r)\sigma^2 = (2/8)\sigma^2$, and Efficiency factor, $E = 3/4$. This shows that efficiency factor is very high.

5. Variance balanced design through augmenting n (≥ 1) blocks

In this section, variance balanced designs are obtained through balanced incomplete block design by augmenting one and more than one blocks, such that each augmented block contains each of the v treatments. The resulting design is an un - equal replicated and un - equal blocks sizes variance balanced design.

Theorem – 5.1 Let N be the incidence matrix of a balanced incomplete block design with parameters v, b, r, k and λ. Let n blocks are added with the blocks of the given balanced incomplete block design. The incidence matrix N_1 defined as

$$N_1 = \left[(N)_{v\,x\,b}\mathbf{1}_{v\,x\,1}\right]$$

gives variance balanced design with parameters $v_1 = v$, $b_1 = b + n$, $r_1 = \{(r + n), b\}$, $k_1 = \{k, v\}$, where, N_1 is the incidence matrix of Variance balanced design.

Proof: Consider a balanced incomplete block design with parameters v, b, r, k and λ, whose incidence matrix is denoted by N. Next n more blocks are augmented, hence, for resulting design, $v_1 = v$, $b_1 = b + n$, $r_1 = (r + n)$, $k_1 = \{k, v\}$. $C_{im} = \frac{\lambda}{k} + \frac{n}{v} = \frac{\lambda v + nk}{vk}$.

Again, a block design Is said to variance balanced design, if C matrix satisfies, $C = \theta\,(I_v - E_{vv}/v)$, where, θ is non zero eigen value of C matrix with multiplicity $(v - 1)$ and $C = \mathrm{diag}(r_1, r_2, \ldots, r_v) - N\,K^{-1}N'$.

$$C = \begin{bmatrix} r+n & 0 & .. & 0 \\ 0 & r+n & .. & 0 \\ : & : & : & : \\ 0 & 0 & .. & r+n \end{bmatrix} - \begin{bmatrix} \mathrm{k(b+n)} & \lambda\mathrm{v}+\mathrm{nk} & .. & \lambda\mathrm{v}+\mathrm{nk} \\ \lambda\mathrm{v}+\mathrm{nk} & \mathrm{k(b+n)} & .. & \lambda\mathrm{v}+\mathrm{nk} \\ : & : & .. & : \\ \lambda\mathrm{v}+\mathrm{nk} & \lambda\mathrm{v}+\mathrm{nk} & .. & \mathrm{k(b+n)} \end{bmatrix} /\mathrm{vk}$$

Diagonal elements are $[\mathrm{vk(r + n) - k(b + n)}]/\mathrm{vk}$, and off diagonal elements are $-\frac{(\lambda\ \mathrm{v+n\ k})}{\mathrm{vk}}$. After simplification, C reduces to C =

$$\begin{bmatrix} \mathrm{vk(r+n)} - k\,(b+n) & -(\lambda\mathrm{v}+\mathrm{nk}) & .. & -(\lambda\mathrm{v}+\mathrm{nk}) \\ -(\lambda\mathrm{v}+\mathrm{nk}) & \mathrm{vk(r+n)} - k\,(b+n) & .. & -(\lambda\mathrm{v}+\mathrm{nk}) \\ : & : & .. & : \\ -(\lambda\mathrm{v}+\mathrm{nk}) & -(\lambda\mathrm{v}+\mathrm{nk}) & .. & \mathrm{vk(r+n)} - k\,(b+n) \end{bmatrix} /\mathrm{vk}$$

$$\text{Finally, } C = \frac{(\lambda\mathrm{v}+\mathrm{nk})}{k}\left[\mathrm{I_v} - \frac{\mathrm{E_{vv}}}{\mathrm{v}}\right] \tag{4}$$

where, $\theta = \frac{(\lambda\mathrm{v+nk})}{k}$ is the non zero eigen value of C matrix with multiplicity $(\mathrm{v} - 1)$. Eq. (4) satisfy the condition of variance balanced design. Hence, this is equi replicated and two unequal block sizes variance balanced design with parameters $\mathrm{v_1 = v}$, $\mathrm{b_1 = b + n}$, $\mathrm{r_1 = (r + n)}$, $\mathrm{k_1 = \{k, v\}}$.

5.1 Efficiency factor

The efficiency factor of a variance balanced design is defined as

$$\mathrm{E} = \frac{\mathrm{Var}\left(\widehat{t_i} - \widehat{t_m}\right)_{RBD}}{\mathrm{Var}\left(\widehat{t_i} - \widehat{t_m}\right)_{VB}}, \text{ where, } \mathrm{Var}\left(\widehat{t_i} - \widehat{t_m}\right)_{RBD} = (2/\mathrm{r})\,\sigma^2 = \frac{2}{r+n}\,\sigma^2,$$

$$\mathrm{Var}\left(\widehat{t_i} - \widehat{t_m}\right)_{VB} = (2/\theta)\sigma^2 = \frac{2}{\frac{(\lambda\mathrm{v+nk})}{k}}\sigma^2 = \frac{2k}{(\lambda\mathrm{v}+\mathrm{nk})} \text{ and } \mathrm{E} = \frac{(\lambda\mathrm{v}+\mathrm{nk})}{k(r+n)}.$$

Example – 5.1 Construct a variance balanced design with parameters $\mathrm{v_1 = 9}$, $\mathrm{b_1 = 15}$, $\mathrm{r_1 = 7}$, $\mathrm{k_1 = \{3, 9\}}$ from a balanced incomplete block design having parameters $\mathrm{v = 9}$, $\mathrm{b = 12}$, $\mathrm{r = 4}$, $\mathrm{k = 3}$, $\lambda = 1$.

Blocks of the balanced incomplete block design are (1 2 3), (4 5 6), (7 8 9), (1 4 7), (2 5 8), (3 6 9), (1 6 8), (2 4 9), (3 5 7), (1 5 9), (2 6 7), (3 4 8). Let n = 3.

Using Theorem – 5.1, incidence matrix of the variance balanced design is given as.

$$\begin{matrix} & 1 & 0 & 0 & 1 & 0 & 0 & 1 & 0 & 0 & 1 & 0 & 0 & 1 & 1 & 1. \\ & 1 & 0 & 0 & 0 & 1 & 0 & 0 & 1 & 0 & 0 & 1 & 0 & 1 & 1 & 1 \\ \mathrm{N_1} = & 1 & 0 & 0 & 0 & 0 & 1 & 0 & 0 & 1 & 0 & 0 & 1 & 1 & 1 & 1 \\ & 0 & 1 & 0 & 1 & 0 & 0 & 0 & 1 & 0 & 0 & 0 & 1 & 1 & 1 & 1 \\ & 0 & 1 & 0 & 0 & 1 & 0 & 0 & 0 & 1 & 1 & 0 & 0 & 1 & 1 & 1 \\ & 0 & 1 & 0 & 0 & 0 & 1 & 1 & 0 & 0 & 0 & 1 & 0 & 1 & 1 & 1 \\ & 0 & 0 & 1 & 1 & 0 & 0 & 0 & 0 & 1 & 0 & 1 & 0 & 1 & 1 & 1 \\ & 0 & 0 & 1 & 0 & 1 & 0 & 1 & 0 & 0 & 0 & 0 & 1 & 1 & 1 & 1 \\ & 0 & 0 & 1 & 0 & 0 & 1 & 0 & 1 & 0 & 1 & 0 & 0 & 1 & 1 & 1 \end{matrix}$$

$\mathrm{N_1}$ gives the incidence matrix of an equi replicated and un - equal block sizes variance balanced design with parameters $\mathrm{v_1 = 9}$, $\mathrm{b_1 = 15}$, $\mathrm{r_1 = 7}$, $\mathrm{k_1 = \{3, 9\}}$ $\mathrm{C_{im} = 2/3}$ and information matrix,

$$C = \begin{bmatrix} 7 & 0 & .. & 0 \\ 0 & 7 & .. & 0 \\ : & : & : & : \\ 0 & 0 & .. & 7 \end{bmatrix} - \begin{bmatrix} 45 & 18 & .. & 18 \\ 18 & 45 & .. & 18 \\ : & : & .. & : \\ 18 & 18 & .. & 45 \end{bmatrix} /27$$

After simplification, C reduces to $C = \begin{bmatrix} 144 & -18 & .. & -18 \\ -18 & 144 & .. & -18 \\ : & : & .. & : \\ -18 & -18 & .. & 144 \end{bmatrix} /27$

Finally,

$$C = \frac{162}{27}\left[I_9 - \frac{E_{99}}{9}\right] = \frac{18}{3}\left[I_9 - \frac{E_{99}}{9}\right] = \theta\left[I_9 - \frac{E_{99}}{9}\right] \qquad (5)$$

Where, $\theta = 6$ is the non zero eigen value of C matrix with multiplicity 8. Hence, it is a variance balance design with $\widehat{t_i} = (1/\theta)\ Q_i = (1/6)\ Q_i$, $\mathrm{Var}(\widehat{t_i} - \widehat{t_m})_{VB} = (2/\theta)\sigma^2 = (2/6)\ \sigma^2$. $\mathrm{Var}(\widehat{t_i} - \widehat{t_m})_{RBD} = (2/r)\ \sigma^2 = (2/7)\ \sigma^2$ and Efficiency factor, E = 6/7. This shows that efficiency factor is very high.

6. Variance balanced design with (v + 1) treatments

Variance balanced design with (v + 1) treatments is constructed by reinforcing one treatment in each block of a balanced incomplete block design.

6.1 Variance balanced designs with (v + 1) treatments from a series of balanced incomplete block design with parameters v, $b = v_{C_2}$, r = $v - 1_{C_{2-1}}$, k = 2 and $\lambda = 1$

In this section, method of the construction of variance balanced design with (v + 1) treatments is discussed. Variance balanced design with (v + 1) treatments can always be constructed through a balanced incomplete block design by reinforcing one treatment and augmenting n blocks. Let the parameters of a balanced incomplete block design are v, $b = v_{C_2}$, r = $v - 1_{C_{2-1}}$, k = 2 and $\lambda = 1$, provided v (r – 1) must be divisible by (k + 1) = 3. This is shown in Theorem – 6.1.

Theorem – 6.1: Let the parameters of an unreduced balanced incomplete block design are v, $b = v_{C_2}$, r = $v - 1_{C_{2-1}}$, k = 2 and $\lambda = 1$, whose incidence matrix is denoted by N. Let balanced incomplete block design is reinforced by one treatment up to b blocks and augmented with n blocks, such that each block contains each of the v treatments, where, n = 1, 2, ... The incidence matrix N_1 defined by

$$N_1 = \begin{bmatrix} N_{v \times b} & E_{v \times n} \\ \mathbf{1}_{1 \times b} & \mathbf{0}_{1 \times n} \end{bmatrix}$$

gives the incidence matrix of a Variance balanced design with parameters $v_1 = (v + 1)$, $b_1 = b + n$, $r_1 = \{r + n, b\}$, k = {3, v}, where, $E_{v \times n}$ is a matrix of v rows and n columns with elements as 1, **1** is a vector of one row and b columns, **0** is a vector of one row and n columns, provided n = $\frac{v\,(r-1)}{3}$, n being integer.

Proof: Let us consider a unreduced balanced incomplete block designs with parameters $b = v_{C_2}$, r = $v - 1_{C_{2-1}}$, k = 2 and $\lambda = 1$, provided v is divisible by k. This series of balanced incomplete block design is reinforced by one more treatment and augmented by n blocks such that $(v + 1)^{th}$ treatment appears in each of the b blocks and each of the n more blocks contains each of the v treatment once and only once, hence, $v_1 = v + 1$, $b_1 = b + n$, $r_1 = \{r + n, b\}$, $k = \{3, v\}$ become the parameters of the resulting variance balanced design. Let us check the C_{im} ($i \neq m = 1$ to v) value for each pair of treatments, C_{im} value for any pair of treatments among v treatments is computed as

$$C_{im} = \frac{1}{3} + \frac{n}{v} \tag{6}$$

Again,

$$C_{im} = \frac{r}{3}, i = 1, \dots, v, \text{ and } m = v + 1 \tag{7}$$

For variance balanced design, C_{im} for each pair of treatment must be same and hence, from (6) and (7), $\frac{1}{3} + \frac{n}{v} = \frac{r}{3}$, or, $\frac{n}{v} = \frac{(r-1)}{3}$, hence, $n = \frac{v(r-1)}{3}$.

Again, a block design Is said to variance balanced design, if C matrix satisfies,

$C = \theta\,(I_v - E_{vv}/v)$, where, θ is non zero eigen value of C matrix with multiplicity $(v - 1)$ and $C = \text{diag}\,(r_1, r_2, \dots, r_v) - N\,K^{-1}N'$.

$$C = \begin{bmatrix} r+n & 0 & .. & 0 \\ 0 & r+n & .. & 0 \\ : & : & : & : \\ 0 & 0 & .. & b \end{bmatrix} - \begin{bmatrix} vr+3n & v+3n & .. & v+3n \\ v+3n & vr+3n & .. & v+3n \\ : & : & .. & : \\ v+3n & v+3n & .. & bv \end{bmatrix} /3v$$

Diagonal elements are (i) $[3v(r + n) - (vr + 3n]/3v$ and (ii) 2bv, and off diagonal elements are - $\frac{(v+3n\,)}{3v}$. After simplification, C reduces to

$$C = \begin{bmatrix} 3v(r+n)\text{–}(vr+3n) & -(v+3n) & .. & -(v+3n) \\ -(v+3n) & 2v(r+n)\text{–}(vr+3n) & .. & -(v+3n) \\ : & : & .. & : \\ -(v+3n) & -(v+3n) & .. & 2bv \end{bmatrix} /3v$$

For variance balanced design, all the diagonal elements must be same and hence, $[3v(r + n) - (vr + 3n] = 2bv$. This shows that one can use either of diagonal element. In this section, we use $[3v(r + n) - (vr + 3n]$ as a diagonal element.

Finally,

$$C = \frac{(3n + 2r + 1)}{3}\left[I_v - \frac{E_{vv}}{v}\right] \tag{8}$$

Where, $\theta = \frac{(3n+2r+1)}{3}$ is the non zero eigen value of C matrix with multiplicity v. Eq. (8) satisfy the condition of variance balanced design. Hence, this is an unequal replicated and unequal block sizes variance balanced design with parameters $v_1 = v$, $b_1 = b + n$, $r_1 = \{(r + n), b\}$, $k_1 = \{3, v\}$.

6.2 Efficiency factor

Since the resulting variance balanced design has two unequal replications and hence, there are two efficiency factors. The efficiency factor of a variance balanced design is defined as.

$E_1 = \frac{Var(\widehat{t_i - t_m})_{RBD1}}{Var(\widehat{t_i - t_m})_{VB}}$, $Var(\widehat{t_i} - \widehat{t_m})_{RBD1} = (2/r)\,\sigma^2 = \frac{2}{r+n}\sigma^2$, where, t_i and t_m are any two treatments among v treatments, that is, $i \neq m = 1$ to v. $Var(\widehat{t_i} - \widehat{t_m})_{RBD2} = (\frac{1}{(r+n)} + \frac{1}{b})\,\sigma^2$, where, $i \neq m = 1$ to v and m = (v + 1).

$$Var(\widehat{t_i} - \widehat{t_m})_{VB} = (2/\theta)\sigma^2 = \frac{2}{\frac{(3n+2r+1)}{3}}\sigma^2 = \frac{6}{(3n+2r+1)}.$$

$$E_1 = \frac{(3n+2r+1)}{3(r+n)}. \text{ Again, } E_2 = \frac{Var(\widehat{t_i} - \widehat{t_m})_{RBD2}}{Var(\widehat{t_i} - \widehat{t_m})_{VB}} = \frac{(b+r+n)(3n+2r+1)}{6b(r+n)}$$

Example – 6.1: Construct a variance balanced design with parameters $v_1 = 6$, $b_1 = 15$, $r_1 = \{9, 10\}$, $k_1 = \{3, 5\}$ from a balanced incomplete block design having parameters v = 5, b = 10, r = 4, k = 2, $\lambda = 1$.

Blocks of the balanced incomplete block design are.

(1 2), (1 3), (1 4), (1 5), (2 3), (2 4), (2 5), (3 4), (3 5), (4 5). Let $n = \frac{v(r-1)}{3} = 5$. Hence, five blocks are augmented.

Using Theorem –6.1, incidence matrix of the variance balanced design is given as.

$$N_1 = \begin{matrix} 1&1&1&1&0&0&0&0&0&0&1&1&1&1&1 \\ 1&0&0&0&1&1&1&0&0&0&1&1&1&1&1 \\ 0&1&0&0&1&0&0&1&1&0&1&1&1&1&1 \\ 0&0&1&0&0&1&0&1&0&1&1&1&1&1&1 \\ 0&0&0&1&0&0&1&0&1&1&1&1&1&1&1 \\ 1&1&1&1&1&1&1&1&1&1&0&0&0&0&0 \end{matrix}$$

N_1 gives the incidence matrix of an unequal replicated and unequal block sizes variance balanced design with parameters $v_1 = 6$, $b_1 = 15$, $r_1 = \{9, 10\}$, $k_1 = \{3, 5\}$, $C_{im} = 20/15$ and information matrix,

$$C = \begin{bmatrix} 9&0&0&0&0&0 \\ 0&9&0&0&0&0 \\ 0&0&9&0&0&0 \\ 0&0&0&9&0&0 \\ 0&0&0&0&9&0 \\ 0&0&0&0&0&10 \end{bmatrix} - \begin{bmatrix} 35&20&20&20&20&20 \\ 20&35&20&20&20&20 \\ 20&20&35&20&20&20 \\ 20&20&20&35&20&20 \\ 20&20&20&20&35&20 \\ 20&20&20&20&20&50 \end{bmatrix} /15$$

After simplification, C reduces to

$$C = \begin{bmatrix} 100&-20&-20&-20&-20&-20 \\ -20&100&-20&-20&-20&-20 \\ -20&-20&100&-20&-20&-20 \\ -20&-20&-20&100&-20&-20 \\ -20&-20&-20&-20&100&-20 \\ -20&-20&-20&-20&-20&100 \end{bmatrix} /15$$

$$\text{Finally, } C = \frac{120}{15}\left[I_6 - \frac{66}{6}\right] = 8\left[I_9 - \frac{E_{99}}{9}\right] \quad = \theta\left[I_6 - \frac{E_{66}}{6}\right] \qquad (9)$$

Where, $\theta = 8$ is the non zero eigen value of C matrix with multiplicity 5. Hence, it is a variance balanced design with $\widehat{t_i} = (1/\theta)\, Q_i = (1/8)\, Q_i$,

$\text{Var}(\widehat{t_i} - \widehat{t_m})_{VB} = (2/\theta)\sigma^2 = (2/8)\,\sigma^2$, $\text{Var}(\widehat{t_i} - \widehat{t_m})_{RBD1} = (2/r)\,\sigma^2 = (2/9)\,\sigma^2$ and $\text{Var}(\widehat{t_i} - \widehat{t_m})_{RBD2} = (\frac{1}{r+n} + \frac{1}{b})\,\sigma^2 = (\frac{1}{9} + \frac{1}{10})\,\sigma^2 = (19/90)\,\sigma^2$ with.

Efficiency factor, $E_1 = 8/9$ and $E_2 = 38/45$. This shows that efficiency factor is very high.

6.3 Variance balanced designs with (v + 1) treatments from a series of balanced incomplete block design with parameters v, $b = v_{C_k}$, r = $v - 1_{C_{k-1}}$, k and $\lambda = v - 2_{C_{k-2}}$

In Section 6.1, method of the construction of variance balanced design with (v + 1) treatments is discussed with block sizes k = 2. In this section, we have extended the method of construction of variance balanced designs with (v + 1) treatments through a balanced incomplete block design for any value of k by reinforcing one treatment and augmenting n blocks. Let the parameters of a balanced incomplete block design are v, $b = v_{C_k}$, $r = v - 1_{C_{k-1}}$, k and $\lambda = v - 2_{C_{k-2}}$, provided, $v(r - \lambda)$ must be divisible by (k + 1). This is shown in Theorem – 6.2.

Theorem – 6.2 Let the parameters of a balanced incomplete block design are v, $b = v_{C_k}$, $r = v - 1_{C_{k-1}}$, k and $\lambda = v - 2_{C_{k-2}}$. whose incidence matrix is denoted by N. Let balanced incomplete block design is reinforced by one treatment up to b blocks and augmented with n blocks, such that each block contains each of the v treatments, where, n = 1, 2, ... The incidence matrix N_1 defined by

$$N_1 = \begin{bmatrix} N_{v \times b} & E_{v \times n} \\ \mathbf{1}_{1 \times b} & \mathbf{0}_{1 \times n} \end{bmatrix}$$

gives the incidence matrix of a Variance balanced design with parameters $v_1 = (v + 1)$, $b_1 = b + n$, $r_1 = \{r + n, b\}$, $k_1 = \{(k + 1), v\}$, where, $E_{v \times n}$ is a matrix of v rows and n columns with elements as 1, **1** is a vector of one row and b columns, **0** is a vector of one row and n columns, provided $n = \frac{v\,(r-\lambda)}{(k+1)}$, n being integers.

Proof: Let us consider a series of balanced incomplete block designs with parameters v, $b = v_{C_k}$, $r = v - 1_{C_{k-1}}$, k and $\lambda = v - 2_{C_{k-2}}$, provided n is divisible by $\frac{v\,(r-\lambda)}{(k+1)}$. This series of balanced incomplete block design is reinforced by one more treatment and augmented by n blocks such that $(v + 1)^{th}$ treatment appears in each of the b blocks and each of the n more blocks contains each of the v treatment once and only once, hence, $v_1 = v + 1$, $b_1 = b + n$, $r_1 = \{r + n, b\}$, $k_1 = \{(k + 1), v\}$ are the parameters of the resulting variance balanced design. Let us check the C_{im} ($i \neq m = 1$ to v) value for each pair of treatments. C_{im} value for any pair of treatments among v treatments is computed as

$$C_{im} = \frac{1}{(k+1)} + \frac{n}{v} \tag{10}$$

$$\text{Again, } C_{im} = \frac{r}{(k+1)}, i = 1, \ldots, v \text{ and } m = v + 1 \tag{11}$$

For variance balanced design, C_{im} for each pair of treatment, must be same and hence, from (10) and (11), $\frac{1}{(k+1)} + \frac{n}{v} = \frac{r}{(k+1)}$, or, $\frac{n}{v} = \frac{(r-\lambda)}{(k+1)}$, hence, $n = \frac{v(r-\lambda)}{(k+1)}$.

Again, a block design Is said to variance balanced design, if C matrix satisfies, $C = \theta\,(I_v - E_{vv}/v)$, where, θ is non zero eigen value of C matrix with multiplicity $(v - 1)$ and $C = \text{diag}(r_1, r_2, \ldots, r_v) - N\,K^{-1}N'$.

$$C = \begin{bmatrix} r+n & 0 & .. & 0 \\ 0 & r+n & .. & 0 \\ : & : & : & : \\ 0 & 0 & .. & b \end{bmatrix} - \begin{bmatrix} vr+n(k+1) & \lambda v+n(k+1) & .. & \lambda v+n(k+1) \\ \lambda v+n(k+1) & vr+n(k+1) & .. & \lambda v+n(k+1) \\ : & : & .. & : \\ \lambda v+n(k+1) & \lambda v+n(k+1) & .. & bv \end{bmatrix} /v(k+1)$$

Diagonal elements are (i) $[v(k + 1)(r + n) - (vr + n(k + 1)]/v(k + 1)$ and (ii) kbv and off diagonal elements are - $\frac{(\lambda v+n(k+1)\ )}{v(k+1)}$, . After simplification, C reduces to

$$C = \begin{bmatrix} vk(n+r)+n(v-k-1) & -(\lambda v+n(k+1)) & .. & -(\lambda v+n(k+1)) \\ -(\lambda v+n(k+1)) & vk(n+r)+n(v-k-1) & .. & -(v+n(k+1)) \\ : & : & .. & : \\ -(\lambda v+n(k+1)) & -(\lambda v+n(k+1)) & .. & kbv \end{bmatrix} /v(k+1)$$

For variance balance design, all the diagonal elements must be same and hence,$vk(n + r) + n(v - k - 1) = kbv$. This shows that we can use either of diagonal element. In this section, we used$vk(n + r) + n(v - k - 1)$ as a diagonal element.

$$\text{Finally, } C = \frac{(n(k+1)+kr+\lambda)}{(k+1)}\left[I_v - \frac{E_{vv}}{v}\right] \tag{12}$$

Where, $\theta = \frac{[n(k+1)+kr+\lambda]}{(k+1)}$ is the non zero eigen value of C matrix with multiplicity v. Eq. (12) satisfy the condition of variance balanced design. Hence, this is an unequal replicated and unequal block sizes variance balanced design with parameters $v_1 = v$, $b_1 = b + n$, $r_1 = \{(r + n), b\}$, $k_1 = \{3(k + 1)\ v\}$.

6.3.1 Efficiency factor of this variance balanced design

Since the resulting variance balanced design is a two unequal replicated design and hence, there are two efficiency factors. The efficiency factor of a variance balanced design is defined as.

$E_1 = \frac{Var\left(\widehat{t_i - t_m}\right)_{RBD1}}{Var\left(\widehat{t_i - t_m}\right)_{VB}}$, $Var\left(\widehat{t_i} - \widehat{t_m}\right)_{RBD1} = (2/r)\ \sigma^2 = \frac{2}{r+n}\sigma^2$, where t_i and t_m are any two treatments among v treatments, $i \neq m = 1$ to v.$Var\left(\widehat{t_i} - \widehat{t_m}\right)_{RBD2} = (\frac{1}{(r+n)} + \frac{1}{b})\ \sigma^2$, where, $i \neq m = 1$ to v and $m = (v + 1)$.

$$Var\left(\widehat{t_i} - \widehat{t_m}\right)_{VB} = (2/\theta)\sigma^2 = \frac{2}{\frac{[n(k+1)+kr+\lambda]}{(k+1)}}\sigma^2 = \frac{2(k+1)}{[n(k+1)+kr+\lambda]}.$$

$$E_1 = \frac{[n(k+1)+kr+\lambda]}{(r+n)(k+1)} \text{ and } E_2 = \frac{[n(k+1)+kr+\lambda](b+r+n)}{2b(r+n)(k+1)}$$

Example – 6.2: Construct a variance balanced design with parameters $v_1 = 6$, $b_1 = 15$, $r_1 = \{9, 10\}$, $k_1 = \{3, 5\}$ from a balanced incomplete block design having parameters $v = 6$, $b = 20$, $r = 10$, $k = 3$, $\lambda = 4$.

Blocks of the balanced incomplete block design are.

(1 2 3), (1 2 4), (1 2 5), (1 2 6), (1 3 4), (1 3 5), (1 3 6), (1 4 5), (1 4 6), (1 5 6), (2 3 4), (2 3 5), (2 3 6), (2 4 5), (2 4 6), (2 5 6), (3 4 5), (3 4 6), (3 5 6), (4 5 6). Let $n = \frac{v(r-\lambda)}{(k+1)} = 9$. Hence, nine blocks are augmented. Using Theorem – 6.2, incidence matrix of the variance balanced design is given as.

$$
\begin{array}{l}
\quad\;\; 1\ 1\ 1\ 1\ 1\ 1\ 1\ 1\ 1\ 1\ 0\ 0\ 0\ 0\ 0\ 0\ 0\ 0\ 0\ 0\ 1\ 1\ 1\ 1\ 1\ 1\ 1\ 1\ 1. \\
\quad\;\; 1\ 1\ 1\ 1\ 0\ 0\ 0\ 0\ 0\ 0\ 1\ 1\ 1\ 1\ 1\ 1\ 0\ 0\ 0\ 0\ 1\ 1\ 1\ 1\ 1\ 1\ 1\ 1\ 1. \\
N_1 = 1\ 0\ 0\ 0\ 1\ 1\ 1\ 0\ 0\ 0\ 1\ 1\ 1\ 0\ 0\ 0\ 1\ 1\ 1\ 0\ 1\ 1\ 1\ 1\ 1\ 1\ 1\ 1\ 1. \\
\quad\;\; 0\ 1\ 0\ 0\ 1\ 0\ 0\ 1\ 1\ 0\ 1\ 0\ 0\ 1\ 1\ 0\ 1\ 1\ 0\ 1\ 1\ 1\ 1\ 1\ 1\ 1\ 1\ 1\ 1. \\
\quad\;\; 0\ 0\ 1\ 0\ 0\ 1\ 0\ 1\ 0\ 1\ 0\ 1\ 0\ 1\ 0\ 1\ 1\ 0\ 1\ 1\ 1\ 1\ 1\ 1\ 1\ 1\ 1\ 1\ 1. \\
\quad\;\; 0\ 0\ 0\ 1\ 0\ 0\ 1\ 0\ 1\ 1\ 0\ 0\ 1\ 0\ 1\ 1\ 0\ 1\ 1\ 1\ 1\ 1\ 1\ 1\ 1\ 1\ 1\ 1\ 1. \\
\quad\;\; 1\ 1\ 1\ 1\ 1\ 1\ 1\ 1\ 1\ 1\ 1\ 1\ 1\ 1\ 1\ 1\ 1\ 1\ 1\ 1\ 0\ 0\ 0\ 0\ 0\ 0\ 0\ 0\ 0.
\end{array}
$$

N_1 gives the incidence matrix of an unequal replicated and un - equal block sizes variance balanced design with parameters $v_1 = 7$, $b_1 = 29$, $r_1 = \{19, 20\}$, $k_1 = \{4, 6\}$; $C_{im} = 10/4$ and information matrix,

$$C = \begin{bmatrix} 19 & 0 & .. & 0 \\ 0 & 19 & .. & 0 \\ : & : & : & : \\ 0 & 0 & .. & 20 \end{bmatrix} - \begin{bmatrix} 48 & 30 & .. & 30 \\ 30 & 48 & .. & 30 \\ : & : & .. & : \\ 30 & 30 & .. & 60 \end{bmatrix} /12$$

After simplification, C reduces to $C = \begin{bmatrix} 180 & -30 & & -30 \\ -30 & 180 & .. & -30 \\ : & : & : & : \\ -30 & -30 & .. & 180 \end{bmatrix} /12$

$$\text{Finally, } C = \frac{210}{12}\left[I_7 - \frac{E_{77}}{7}\right] = \frac{35}{2}\left[I_7 - \frac{E_{77}}{7}\right] \qquad = \theta\left[I_7 - \frac{E_{77}}{7}\right] \tag{13}$$

Where, $\theta = 35/2$ is the non zero eigen value of C matrix with multiplicity 6. Hence, it is a variance balanced design with$\widehat{t_i} = (1/\theta)\ Q_i = (2/35)\ Q_i$, $Var\left(\widehat{t_i} - \widehat{t_m}\right)_{VB} = (2/\theta)\sigma^2 = (4/35)\ \sigma^2$,$Var\left(\widehat{t_i} - \widehat{t_m}\right)_{RBD1} = (2/r)\ \sigma^2 = (2/19)\ \sigma^2$ and $Var\left(\widehat{t_i} - \widehat{t_m}\right)_{RBD2} = \left(\frac{1}{r+n} + \frac{1}{b}\right)\sigma^2 = \left(\frac{1}{9} + \frac{1}{10}\right)\sigma^2 = (39/380)\ \sigma^2$ with efficiency factor, $E_1 = 35/38$ and $E_2 = 273/304$. This shows that efficiency factor is very high.

Nonexistence of variance balanced design by reinforcing (v + t) treatments.

Variance balanced design cannot be constructed from a balanced incomplete block design by reinforcing (v + t) treatments, t = 1, 2, ...

Because for variance balanced design C_{im} must be same for each pair of treatments. In this case $C_{im} = \frac{\lambda}{k+1}$, $i \neq m = 1$ to v and $C_{i\ (v\ +\ 1)} = \frac{r}{k+1}$, where, t = 1. As per the condition of variance balanced design, $\frac{r}{k+1} = \frac{\lambda}{k+1}$, which is not possible as $(r - \lambda) > 0$. If two treatments are reinforced, then $\frac{r}{k+1} = \frac{\lambda}{k+1} = \frac{b}{k+1}$ must holds true, but $b \neq r \neq \lambda$.

7. Variance balanced design using 2^n symmetrical factorial experiments

This section discusses the construction of variance balanced design using 2^n symmetrical factorial experiment.

Theorem: 7.1 Let us consider a 2^n factorial experiment. By deleting the control treatment and all the main effects, equi-replicated and unequal block sizes variance balanced design is obtained with parameters v = n, b = $2^n - n - 1$, r = 2^{n-1}, k = {2, 3, 4 ... , n}.

Proof: consider a 2^n treatment combination of a 2^n factorial experiment. Delete its control treatment and all main effects. Consider each treatment combination as one block. So, we have $(2^n - 1 - n)$ blocks with v (= n) treatments, as factors are considered as treatments. Consider this matrix as an incidence matrix of a block design, whose all elements are either zero or 1 only. So, the design is binary.

Since each treatment is repeated $(2^{n-1} - 1)$ times, design is equi replicated and unequal block sizes with k = {2, 3, ... , n}. Let the incidence matrix of the block design is given by

$$N = \begin{bmatrix} 0 & 1 & 1 & 1 & \vdots & 1 \\ 1 & 0 & 0 & 1 & \vdots & 1 \\ 1 & 1 & 1 & 0 & \vdots & 1 \\ \vdots & \vdots & \vdots & \vdots & \vdots & 1 \\ 0 & 0 & 1 & 1 & \vdots & 1. \end{bmatrix}$$

The incidence matrix of the block design is a variance balanced, if the C matrix of the block design satisfy $C = \theta\left[I_v - \frac{1}{v}E_{VV}\right]$, where, θ is non – zero eigen value of C matrix.

$$C = \begin{bmatrix} 2^{n-1} - 1 & 0 & \dots & 0 \\ 0 & 2^{n-1} - 1 & \dots & 0 \\ 0 & 0 & \dots & 2^{n-1} - 1 \end{bmatrix} - \begin{bmatrix} Y & X & \dots & X \\ X & Y & \dots & X \\ \vdots & \vdots & \vdots & \vdots \\ X & X & \dots & Y \end{bmatrix}$$

where, $Y = \frac{\binom{n-1}{1}}{2} + \frac{\binom{n-1}{2}}{3} + \frac{\binom{n-1}{3}}{4} + \dots + \frac{1}{n}$, and

$X = \frac{1}{2} + \frac{n-2}{3} + \frac{n-3}{4} + \dots + \frac{n-(n-1)}{n}$ and $r = 2^{n-1} - 1$

After simplification, C reduces to

$$C = \begin{bmatrix} 2^{n-1} - 1 - Y & -X & \dots & -X \\ -X & 2^{n-1} - 1 - Y & \dots & -X \\ \vdots & \vdots & & \vdots \\ -X & -X & \dots & 2^{n-1} - 1 - Y \end{bmatrix}$$

Finally,

$$C = (2^{n-1} - 1 - Y + X)\left[I_v - \frac{E_{vv}}{v}\right] \tag{14}$$

Where, $\theta = (2^{n-1} - 1 - Y + X)$ is the non-zero eigen value of C matrix with multiplicity (v – 1). Eq. (14) satisfy the condition of variance balanced design. Hence, this is an equal replicated and unequal block sizes variance balanced design with parameters v = n, b = $2^n - n - 1$, r = 2^{n-1} - 1, k = {2, 3, 4 ..., n.}. Efficiency factor, E = $\frac{(2^{n-1}-1-Y+X)}{(2^{n-1}-1)}$.

Example: 7.1. Construct a variance balanced design with parameters v = 4, b = 11, r = 7, k = {2, 3, ..., n.}

Using Theorem 7.1, incidence matrix of the variance balanced design is given by

$$N = \begin{bmatrix} 0\,0\,0\,0\,1\,1\,1\,1\,1\,1\,1 \\ 0\,1\,1\,1\,0\,0\,0\,1\,1\,1\,1 \\ 1\,0\,1\,1\,0\,1\,1\,0\,0\,1\,1 \\ 1\,1\,0\,1\,1\,0\,1\,0\,1\,0\,1 \end{bmatrix}$$

$$C = \begin{bmatrix} 7 & 0 & 0 & 0 \\ 0 & 7 & 0 & 0 \\ 0 & 0 & 0 & 0 \\ 0 & 0 & 0 & 7 \end{bmatrix} - \begin{bmatrix} 33 & 17 & 17 & 17 \\ 17 & 33 & 17 & 17 \\ 17 & 17 & 33 & 17 \\ 17 & 17 & 17 & 33 \end{bmatrix} /12$$

After simplification, C reduces to

$$C = \begin{bmatrix} 51 & -17 & -17 & -17 \\ -17 & 51 & -17 & -17 \\ -17 & -17 & 51 & -17 \\ -17 & -17 & -17 & 51 \end{bmatrix} /12$$

Finally,

$$C = \frac{68}{12}\left[I_4 - \frac{E_{44}}{4}\right] = \frac{17}{3}\left[I_4 - \frac{E_{44}}{4}\right] = \theta\left[I_4 - \frac{44}{4}\right] \qquad (15)$$

Where, θ = 17/3 is the non zero eigen value of C matrix with multiplicity 3. Hence, it is a variance balanced design with $\widehat{t_i}$ = (1/ θ) Q_i = (3/17) Q_i, $Var(\widehat{t_i} - \widehat{t_m})_{VB}$ = (2/θ)σ^2 = (6/17) σ^2, $Var(\widehat{t_i} - \widehat{t_m})_{RBD}$= (2/r) σ^2 = (2/7) σ^2 and efficiency factor, E = 17/21.

This result is due to Ghosh, Sinojia and Ghosh (2018).

8. Construction of variance balanced designs using some incidence matrix

Theorem 8.1: Let I_n denotes the identity matrix of order n, j_n is a column Vector of one, 0_n is the row vectors having all elements zero. An incidence matrix N defined as $N = \begin{bmatrix} j'_n & \mathbf{0}_{1\times\frac{n}{2}} \\ I_n & E_{nxn/2} \end{bmatrix}$

gives the incidence matrix of a Variance balanced designs, with parameters $v = n + 1, b = n + \frac{n}{2}, r = \left\{n, 1 + \frac{n}{2}\right\}$ and $k = \{2, n\}$, where, n is even.

Proof: Proof is obvious.

Example 8.1. Let = 6. So, the incidence matrix N using Theorem 8.1 is given by

$$N = \begin{bmatrix} 1 & 1 & 1 & 1 & 1 & 1 & 0 & 0 & 0 \\ 1 & 0 & 0 & 0 & 0 & 0 & 1 & 1 & 1 \\ 0 & 1 & 0 & 0 & 0 & 0 & 1 & 1 & 1 \\ 0 & 0 & 1 & 0 & 0 & 0 & 1 & 1 & 1 \\ 0 & 0 & 0 & 1 & 0 & 0 & 1 & 1 & 1 \\ 0 & 0 & 0 & 0 & 1 & 0 & 1 & 1 & 1 \\ 0 & 0 & 0 & 0 & 0 & 1 & 1 & 1 & 1 \end{bmatrix}$$

$$C = \begin{bmatrix} 6 & 0 & 0 & 0 & 0 & 0 & 0 \\ 0 & 4 & 0 & 0 & 0 & 0 & 0 \\ 0 & 0 & 4 & 0 & 0 & 0 & 0 \\ 0 & 0 & 0 & 4 & 0 & 0 & 0 \\ 0 & 0 & 0 & 0 & 4 & 0 & 0 \\ 0 & 0 & 0 & 0 & 0 & 4 & 0 \\ 0 & 0 & 0 & 0 & 0 & 0 & 4 \end{bmatrix} - \begin{bmatrix} 6 & 1 & 1 & 1 & 1 & 1 & 1 \\ 1 & 2 & 1 & 1 & 1 & 1 & 1 \\ 1 & 1 & 2 & 1 & 1 & 1 & 1 \\ 1 & 1 & 1 & 2 & 1 & 1 & 1 \\ 1 & 1 & 1 & 1 & 2 & 1 & 1 \\ 1 & 1 & 1 & 1 & 1 & 2 & 1 \\ 1 & 1 & 1 & 1 & 1 & 1 & 2 \end{bmatrix} /2$$

After simplification, C reduces to

$$C = \begin{bmatrix} 6 & -1 & -1 & -1 & -1 & -1 & -1 \\ -1 & 6 & -1 & -1 & -1 & -1 & -1 \\ -1 & -1 & 6 & -1 & -1 & -1 & -1 \\ -1 & -1 & -1 & 6 & -1 & -1 & -16 \\ -1 & -1 & -1 & -1 & 6 & -1 & -1 \\ -1 & -1 & -1 & -1 & -1 & 6 & -1 \\ -1 & -1 & -1 & -1 & -1 & -1 & 6 \end{bmatrix} /2$$

Finally,

$$C = \frac{7}{2}\left[I_7 - \frac{E_{77}}{7}\right] = \frac{7}{2}\left[I_7 - \frac{E_{77}}{7}\right] \qquad = \theta \left[I_7 - \frac{E_{77}}{7}\right] \tag{16}$$

where, $\theta = 7/2$ is the non zero eigen value of C matrix with multiplicity 6. Hence, it is a variance balanced design with parameters $v = 7$, $b = 9$, $r = \{6,4\}$, $k = \{2, 6\}$. $\widehat{t_i} = (1/\theta)\, Q_i = (2/7)\, Q_i$; $\mathrm{Var}(\widehat{t_i} - \widehat{t_m})_{VB} = (2/\theta)\sigma^2 = (4/7)\,\sigma^2$, $\mathrm{Var}(\widehat{t_i} - \widehat{t_m})_{RBD1} = (2/r)\,\sigma^2 = (2/4)\,\sigma^2$, $\mathrm{Var}(\widehat{t_i} - \widehat{t_m})_{RBD2} = (\frac{1}{6} + \frac{1}{4})\,\sigma^2 = (5/12)\,\sigma^2$ with efficiency factor, $E_1 = 7/8$ and efficiency factor, $E_2 = 35/48$.

This result is due to Ghosh, Sinojia and Ghosh (2018).

9. Conclusions

In this chapter, we have constructed Variance balanced designs using balanced incomplete block, group divisible, resolvable semi - regular group divisible, symmetrical factorial and fractional factorial designs. It is observed that efficiency factor of all most all variance balanced design is high. Variance balanced designs constructed in sections 3 to 6 are new and extended methods, while Section 7 and 8, discuss the review work of Ghosh, Sinojia and Ghosh (2018).

Acknowledgements

Author is very much grateful to University Grants Commission, New Delhi, for providing me an opportunity to work as UGC BSR Faculty Fellow. Author is also thankful to referees for suggesting the important ideas in improving the present chapter.

Author details

D.K. Ghosh
Department of Statistics, Saurashtra University, Rajkot, Gujarat, India

*Address all correspondence to: ghosh_dkg@yahoo.com

References

[1] Raghavarao D. Constructions and Combinatorial Problems in Design of Experiments. New York: Wiley; 1971

[2] Chakrabarti MC. On the C-matrix in design of experiments. Journal of the Indian Statistical Association. 1963;**1**: 8-23. 8C23

[3] Pearce SC. Experimenting with blocks of natural size. Biometrics. 1964; **20**(4):699-706. 699C706. DOI: 10.2307/2528123

[4] Kageyama S. Reduction of associate classes for block designs and related combinatorial arrangements. Hiroshima Mathematical Journal. 1974;**4**:527C618

[5] Rao VR. A note on balanced designs. The Annals of Mathematical Statistics. 1958;**29**(1):290-294. 290C294. DOI: 10.1214/aoms/1177706729

[6] Hedayat A, Federer WT. Pairwise and variance balanced incomplete block design. Annals of the Institute of Statistical Mathematics. 1974;**26**(1): 331-338. DOI: 10.1007/BF02479828

[7] Raghavarao D. On balanced unequal block designs. Biometrika. 1962;**49**(3/4):561-562. 561C562. DOI: 10.1093/biomet/49.3-4.561

[8] Puri PD, Nigam AK. Balanced block designs. Communications in Statistics – Theory and Methods. 1977;**6**(12): 1171-1179. 1171C1179. DOI: 10.1080/03610927708827560

[9] John PWM. Balanced designs with unequal number of replications. Annals of Mathematical Statistics. 1964;**35**: 897C899. DOI: 10.1214/aoms/1177703597

[10] Jones B, sinha, K., & Kageyama, S. Further equireplicated variance balanced designs with unequal block sizes. Utilitas Mathematica. 1987;**32**: 5C10

[11] Kageyama S. Construction of balanced block designs. Utilitas Mathematica. 1976;**9**:209C229

[12] Kageyama S. Existence of variance-balanced binary designs with fewer experimental units. Statistics & Probability Letters. 1988;**7**(1):27-28. 27C28. DOI: 10.1016/0167-7152(88) 90083-1

[13] Kulshreshtha AC, Dey A, Saha GM. Balanced designs with unequal replications and unequal block sizes. Annals of Mathematical Statistics. 1972; **43**:1342-1345

[14] Pal S, Pal S. Nonproper variance balanced designs and optimality. Communications in Statistics - Theory and Methods. 1988;**17**:1685C1695. DOI: 10.1080/03610928808829706

[15] Roy J. On the efficiency factor of block designs. Sankhya. 1958;**19**: 181C188

[16] Sinha K. A note on equireplicated balanced block designs from BIB designs. Journal of the Indian Society of Agricultural Statistics. 1988;**42**:150C153

[17] Sinha K. Some new equireplicated balanced block designs. Statistics & Probability Letters. 1989;**8**:89. DOI: 10.1016/0167-7152(89)90089-8

[18] Sinha K, Jones B. Further equireplicated balanced block designs with unequal block sizes. Statistics & Probability Letters. 1988;**6**:229C330

[19] Tyagi BN. On a class of variance balanced block designs. Journal of Statistical Planning and Inference. 1979; **3**:333-336. DOI: 10.1016/0378-3758(79) 90029-6

[20] Khatri CG. A note on variance balanced designs. Journal of Statistical Planning and Inference. 1982;**6**(2):

173-177. 173C177. DOI: 10.1016/0378-3758(82)90086-6

[21] Das MN, Ghosh DK. Balancing incomplete block designs. Sankhyā: The Indian Journal of Statistics, Series B. 1985;**47**(1):67-77. 67C77

[22] Mukerjee R, Kageyama S. On resolvable and affine resolvable variance-balanced designs. Biometrika. 1985;**72**(1):165-172. 165C172. DOI: 10.1093/biomet/72.1.165

[23] Calvin JA. A new class of variance balanced designs. Journal of Statistical Planning and Inference. 1986;**14**(2–3): 251-254. DOI: 10.1016/0378-3758(86)90162-X

[24] Calvin JA, Sinha K. A method of constructing variance balanced designs. Journal of Statistical Planning and Inference. 1989;**23**(1):127-131. DOI: 10.1016/0378-3758(89)90045-1

[25] Agarwal GG, Kumar S. On a class of variance balanced designs associated with GD designs. Calcutta Statistics Association Bulletin. 1984;**33**(3–4): 187-190. 187C190

[26] Ghosh DK. A series of generalized efficiency balanced designs. Gujarat Statistical Review. 1988;**15**(1):33-38. 33C38

[27] Ghosh DK, Karmokar PK. Some series of efficiency balanced designs. Australian & New Zealand Journal of Statistics. 1988;**30**(1):47-51. 47C51

[28] Ghosh DK, Divecha J. Construction of variance balanced designs. TEC.R. No.2, Dept. of Maths. and Stat., Saurashtra Uni. 1989:1-13. 1C13

[29] Ghosh DK, Divecha J, Kageyama S. Equi - replicate variance balanced designs from group divisible designs. Journal of the Japan Statistical Society. 1991;**21**(1):205-209. 205C209. DOI: 10.1111/j.1467-842X.1988.tb00610.x

[30] Ghosh DK, Joshi K. Note on construction of variance balanced designs through group divisible designs. Utilitas Mathematica. 1991;**39**:249-253. 249C253

[31] Ghosh DK, Karmokar PK, Divecha J. On Comparison of ternary efficiency balanced and ternary variance balanced designs. Utilitas Mathematica. 1991;**40**: 25-27

[32] Ghosh DK, Anita S, Kageyama S. Construction of variance-balanced and efficiency balanced ternary block designs. Journal of Japan Statistical Association Soc. 1994;**24**(2):201-208

[33] Ghosh DK, Joshi K. Construction of variance balanced designs through triangular PBIB designs. Calcutta Statistical Association Bulletin. 1995;**45** (1–2):111-118

[34] Ghosh DK, Sangeeta A. On variance balanced designs. Journal of Modern Applied Statistical Methods. 2017;**16**(2): 124-137

[35] Agarwal GG, Kumar S. A note on construction of variance balanced designs. Journal of the Indian Society of Agricultural Statistics. 1985;**37**(2): 181-183. 181C183

[36] Agarwal GG, Kumar S. Construction of balanced ternary designs. Australian and Newziland Journal of Statistics. 1986;**28**(2):251-255. 251C255. DOI: 10.1111/j.1467-842X.1986.tb00605.x

[37] Ghosh DK, Joshi K, Kageyama S. Ternary variance balanced designs through BIB and GD designs. Journal of the Japan Statistical Society. 1993;**23**(1): 75-81. 75C81

[38] Ghosh DK. Robustness of Variance balanced designs against the loss of one block. Indian Journal of Statistics and Applications. 2012;**1**:17-30. 17C30

[39] Ghosh DK, Sinojia CN, Ghosh S. On a class of variance balanced designs.

International Journal of Statistics and Probability. 2018;7(3):112-120

[40] Hedayat A, Stufken JA. Relation between pair wise balanced and variance balanced block designs. Journal of the American Statistical Association. 1989;**84**(407):753-755. DOI: 10.1080/01621459.1989.10478830

[41] Jones RM. On a property of incomplete blocks. Journal of the Royal Statistical Society: Series B. 1959;**21**: 172-179. Available from: http://www.jstor.org/stable/2983939

[42] Gupta SC, Jones B. Equi-replicate balanced block designs with unequal block sizes. Biometrika. 1983;**70**(2): 443-440. 443C440. DOI: 10.1093/biomet/70.2.433

Chapter 9

Estimation of Means of Two Quantitative Sensitive Variables Using Randomized Response Technique

Amod Kumar

Abstract

I propose an improved randomized response model for the simultaneous estimation of population means of two quantitative sensitive variables by using blank card option that make use of one scramble response and another fake response. The properties of the proposed estimator have been analysed. To judge the performance of the proposed model, I have considered a real data set and it is to be pointed out that the proposed model is more efficient in terms of relative efficiencies and privacy protection of respondents as well. Suitable recommendations have been made to the survey practitioners.

Keywords: randomized response technique, two quantitative sensitive variables, estimation of two means\, blank card, privacy protection

1. Introduction

Reliability of data is compromised when sensitive topics on embarrassing or illegal acts such as students taking drugs, drunk driving, abortion, family income, tax evasion etc. are required in direct method of data collection in sample survey. Survey on human population has established the fact that the direct question about sensitive characters often results in either refusal to respond or falsification of the answer. To overcome this difficulty and ensure confidentially of respondents, Warner [1] initiated a technique which is called as randomized response technique (RRT). For estimating π, the population proportion of respondents, a simple random sample of size n respondents selected from the population N with replacement. Each respondent selected in the sample has a random device which consists two statements "I belong to sensitive group A" and "its compliment A^c". The respondent answers of sensitive or non-sensitive questions depending on the outcome of the random device which is unobservable to the sampler. Greenberg et al. [2] adjusted the Warner [1] model with respect to efficiency and respondent's cooperation by suggesting unrelated question randomized response model, where the sensitive question was combined with an unrelated (non-sensitive) question.

Greenberg et al. [3] extended the Greenberg et al. [2] model to estimate the population mean of quantitative sensitive variable, such as income, tax dodging etc. In their model, each respondent selected in the sample with replacement was given

IntechOpen

a random device which presents two outcomes Y and X with probabilities P and (1-P) respectively, where Y is the true quantitative sensitive variable and X is non-sensitive independent variable. Later, Eichhorn and Hayre [4] introduced a new multiplicative randomized response model for estimating the population mean of quantitative sensitive variable.

Under simple random sampling with replacement (SRSWR) scheme, Perri [5] modified Greenberg et al. [3] technique to obtain the estimator of population mean μ_Y by using a blank card option, if a blank card is selected then the respondents are requested to use Greenberg et al. [3] model. In his model, the observed response θ_P is given by:

$$\theta_P = \begin{cases} Y & \text{with probability } P_1 \\ Y & \text{with probability } P_2 \\ \text{Blank Card} & \text{with probability } P_3 \end{cases} \tag{1}$$

Perri [5] proposed an unbiased estimator of the population mean μ_Y

$$\hat{\mu}_P = \frac{\overline{\theta}_P\text{-}\{P_2 + P_3(1\text{-}P)\}\mu_X}{(P_1 + P_3 P)} \tag{2}$$

with variance

$$V(\hat{\mu}_P) = \frac{\sigma^2_{\theta_P}}{n(P_1 + P_3 P)^2} \tag{3}$$

where $\overline{\theta}_P = 1/n\sum_{i=1}^{n}\theta_{Pi}$ and

$$\sigma^2_{\theta_P} = (P_1 + P_3 P)(\sigma^2_Y + \mu^2_Y) + \{P_2 + P_3(1\text{-}P)\}(\sigma^2_X + \mu^2_X)\text{-}[(P_1 + P_3 P)\mu_Y + \{P_2 + P_3(1\text{-}P)\}\mu_X]^2$$

Many different suggestions have been made for the use of these blank cards by various authors including Bhargava and Singh [6], Singh et al. [7], Batool et al. [8], Singh [9] and Singh et al. [10, 11] among others. Furthermore in addition, the theory of randomized response technique to estimate the population parameters of sensitive characteristics was extended by Narjis and Shabbir [12, 13].

Recently, Ahmed et al. [14] have introduced the idea to estimate the means of two quantitative sensitive variables simultaneously by using one scramble response and other face response. Let Y_{1i} and Y_{2i} be the two values of quantitative sensitive variables with means (μ_{Y1}, μ_{Y2}) and variances $(\sigma^2_{Y1}, \sigma^2_{Y2})$ respectively connected with the ith unit in the populationN. The parameters of interest are (μ_{Y1}, μ_{Y2}) which are to be estimated. Each respondent selected in the sample with replacement is asked to produce two fake values of scramble variables S_1 and S_2 from two known distributions. Let S_1 and S_2 be the independent scramble variables with known means (θ_1, θ_2) and variance $(\gamma_{20}, \gamma_{02})$ respectively, which help to maintain the protection of respondents. Ahmed et al. [14] defined the scramble response as:

$$Z_{1i} = S_1 Y_{1i} + S_2 Y_{2i} \tag{4}$$

Each respondent selected in the sample is also requested to draw a card from the deck which consist two types of cards, similar to Warner [1] model but has different type of outcomes. Let P be the probability of cards bearing the statements in the deck, "the selected respondent to report scramble response as S_1" and (1-P) is the probability of cards bearing the statement in the deck, "the selected respondent to report scramble response as S_2". Thus, the second response from the ith respondent given as:

$$Z_i = \begin{cases} S_1 & \text{with probability } P \\ S_2 & \text{with probability } (1\text{-}P) \end{cases} \tag{5}$$

where $P \neq \frac{\theta_1\gamma_{20}}{\theta_1\gamma_{20} + \theta_2\gamma_{02}}$

Ahmed et al. [14] proposed unbiased estimators of population means μ_{Y1} and μ_{Y2} respectively, and are given as:

$$\hat{\mu}_{AY1} = \frac{\left[P\theta_1\theta_2 + (1\text{-}P)\left(\gamma_{02} + \theta_2^2\right)\right]\overline{Z}_1\text{-}\theta_2\overline{Z}_2}{(1\text{-}P)\theta_1\gamma_{02}\text{-}P\theta_2\gamma_{20}} \tag{6}$$

and

$$\hat{\mu}_{AY2} = \frac{\theta_1\overline{Z}_2\text{-}\left[P\left(\gamma_{20} + \theta_1^2\right) + (1\text{-}P)\theta_1\theta_2\right]\overline{Z}_1}{(1\text{-}P)\theta_1\gamma_{02}\text{-}P\theta_2\gamma_{20}} \tag{7}$$

where $\overline{Z}_1 = \frac{1}{n}\sum_{i=1}^{n} Z_{1i}$ and $\overline{Z}_2 = \frac{1}{n}\sum_{i=1}^{n} Z_{2i}$

In follow up of above works and motivated by Ahmed et al. [14], I adopt Perri [5] method and proposes a new improved randomized response model by introducing blank card option for estimation of population means of two quantitative sensitive variables. For example, Y_1 may stand for the respondents' income and Y_2 may stand for the respondents' expenditure, Y_1 denotes the import in millions and Y_2 denotes the export in millions etc. I have demonstrated the efficacious performance of the proposed randomized response model over the Ahmed et al. [14] model along with privacy protection of respondents.

2. Proposed model

In the proposed model, I have considered the similar supposition as it is the case of Ahmed et al. [14] procedure with the modification that the second response of Warner [1] method is replaced with Perri [5] blank card method. Proceeding on the lines of Ahmed et al. [14] as given in their model, the first observed response is given by:

$$Z_{1i} = S_1 Y_{1i} + S_2 Y_{2i} \tag{8}$$

Noted that by mixing of two quantitative sensitive variables with two scramble variables will make more comfortable to respondent about providing information because it make very hard to guess the true value of two quantitative sensitive variables to an interviewer.

Here I differ from the existing randomized response model available in the literature, in that, the second response is replaced with Perri [5] procedure but has different outcomes. Each selected respondent in the sample provided a random device which consists three type of cards bearing the statements (i) green cards with the statement: report scramble variable S_1, (ii) red cards with the statement: report scramble variable S_2 and (iii) yellow card with no statement (blank cards) with probabilities P_1, P_2 and P_3 respectively such that $\sum_{i=1}^{3} P_i = 1$. Thus, the second response Z_{Ai} in the proposed model from i^{th} respondent is given by:

$$Z_{Ai} = \begin{cases} S_1 & \text{with probability } P_1 \\ S_2 & \text{with probability } P_2 \\ \text{Blank card} & \text{with probability } P_3 \end{cases} \tag{9}$$

If a blank card is selected, the respondents are requested to use Ahmed et al. [14] second response. Thus, the second response Z_{Ai} can be rewritten as:

$$Z_{Ai} = \begin{cases} S_1 & \text{with probability } P_1 \\ S_2 & \text{with probability } P_2 \\ \left.\begin{cases} S_1 & \text{with probability } P \\ S_2 & \text{with probability } (1\text{-}P) \end{cases}\right\} & \text{with probability } P_3 \end{cases} \tag{10}$$

where $P \neq \dfrac{(P_2 + P_3)\theta_1\gamma_{02}\text{-}P_2\theta_2\gamma_{20}}{P_3(\theta_1\gamma_{02} + \theta_2\gamma_{20})}$

Taking expectation on both sides of (Eq. (8)), I have

$$E(Z_{1i}) = E(S_1Y_{1i} + S_2Y_{2i}) = \theta_1\mu_{Y1} + \theta_2\mu_{Y2} \tag{11}$$

With the help from Eqs. (8) and (10), I generate a new response Z'_{2i} as:

$$Z'_{2i} = Z_{1i}Z_{Ai}$$
$$= \begin{cases} S_1^2Y_{1i} + S_1S_2Y_{2i} & \text{with probability } P_1 \\ S_1S_2Y_{1i} + S_2^2Y_{2i} & \text{with probability } P_2 \\ \left.\begin{cases} S_1^2Y_{1i} + S_1S_2Y_{2i} & \text{with probability } P \\ S_1S_2Y_{1i} + S_2^2Y_{2i} & \text{with probability } (1\text{-}P) \end{cases}\right\} & \text{with probability } P_3 \end{cases} \tag{12}$$

Taking expectation on both sides of (Eq. (12)), I get

$$\begin{aligned} E(Z'_{2i}) = {} & (P_1 + P_3P)\left[(\gamma_{20} + \theta_1^2)\mu_{Y1} + \theta_1\theta_2\mu_{Y2}\right] \\ & + \{P_2 + P_3(1\text{-}P)\}\left[\theta_1\theta_2\mu_{Y1} + (\gamma_{02} + \theta_2^2)\mu_{Y2}\right] \end{aligned} \tag{13}$$

Using the method of moments on Eqs. (11) and (13), I have:

$$\theta_1\hat{\mu}_{Y1} + \theta_2\hat{\mu}_{Y2} = \frac{1}{n}\sum_{i=1}^{n} Z_{1i} \tag{14}$$

and

$$\begin{aligned} & \left[(P_1 + P_3P)(\gamma_{20} + \theta_1^2) + \{P_2 + P_3(1\text{-}P)\}\theta_1\theta_2\right]\hat{\mu}_{Y1} \\ & + \left[(P_1 + P_3P)\theta_1\theta_2 + \{P_2 + P_3(1\text{-}P)\}(\gamma_{02} + \theta_2^2)\right]\hat{\mu}_{Y2} = \frac{1}{n}\sum_{i=1}^{n} Z'_{2i} \end{aligned} \tag{15}$$

Eqs. (14) and (15) can be rewritten as:

$$\begin{bmatrix} \theta_1, & \theta_2 \\ (P_1 + P_3P)(\gamma_{20} + \theta_1^2) & (P_1 + P_3P)\theta_1\theta_2 \\ +\{P_2 + P_3(1\text{-}P)\}\theta_1\theta_2, & +\{P_2 + P_3(1\text{-}P)\}(\gamma_{02} + \theta_2^2) \end{bmatrix} \begin{bmatrix} \hat{\mu}_{Y1} \\ \hat{\mu}_{Y2} \end{bmatrix} = \begin{bmatrix} \overline{Z}_1 \\ \overline{Z}'_2 \end{bmatrix} \tag{16}$$

Applying Cramer's rule on Eq. (16), I obtain

$$\Delta = \begin{vmatrix} \theta_1, & \theta_2 \\ (P_1+P_3P)\left(\gamma_{20}+\theta_1^2\right)+\{P_2+P_3(1\text{-}P)\}\theta_1\theta_2, & (P_1+P_3P)\theta_1\theta_2+\{P_2+P_3(1\text{-}P)\}\left(\gamma_{02}+\theta_2^2\right) \end{vmatrix}$$

$$= \theta_1\left[(P_1+P_3P)\theta_1\theta_2+\{P_2+P_3(1\text{-}P)\}\left(\gamma_{02}+\theta_2^2\right)\right]$$

$$-\ \theta_2\left[(P_1+P_3P)\left(\gamma_{20}+\theta_1^2\right)+\{P_2+P_3(1\text{-}P)\}\theta_1\theta_2\right]$$

$$= \{P_2+P_3(1\text{-}P)\}\theta_1\gamma_{02}\text{-}(P_1+P_3P)\theta_2\gamma_{20} \tag{17}$$

$$\Delta_1 = \begin{vmatrix} \overline{Z}_1, & \theta_2 \\ \overline{Z}_2', & (P_1+P_3P)\theta_1\theta_2+\{P_2+P_3(1\text{-}P)\}\left(\gamma_{02}+\theta_2^2\right) \end{vmatrix}$$

$$= \left[(P_1+P_3P)\theta_1\theta_2+\{P_2+P_3(1\text{-}P)\}\left(\gamma_{02}+\theta_2^2\right)\right]\overline{Z}_1\text{-}\theta_2\overline{Z}_2' \tag{18}$$

and

$$\Delta_2 = \begin{vmatrix} \theta_1 & \overline{Z}_1 \\ (P_1+P_3P)\left(\gamma_{20}+\theta_1^2\right)+\{P_2+P_3(1\text{-}P)\}\theta_1\theta_2, & \overline{Z}_2' \end{vmatrix}$$

$$= \theta_1\overline{Z}_2'\text{-}\left[(P_1+P_3P)\left(\gamma_{20}+\theta_1^2\right)+\{P_2+P_3(1\text{-}P)\}\theta_1\theta_2\right]\overline{Z}_1 \tag{19}$$

Thus, the estimators of the population mean μ_{Y1} and μ_{Y2} are respectively given by:

$$\hat{\mu}_{Y1} = \frac{\Delta_1}{\Delta} = \frac{\left[(P_1+P_3P)\theta_1\theta_2+\{P_2+P_3(1\text{-}P)\}\left(\gamma_{02}+\theta_2^2\right)\right]\overline{Z}_1\text{-}\theta_2\overline{Z}_2'}{\left[\{P_2+P_3(1\text{-}P)\}\theta_1\gamma_{02}\text{-}(P_1+P_3P)\theta_2\gamma_{20}\right]} \tag{20}$$

and

$$\hat{\mu}_{Y2} = \frac{\Delta_2}{\Delta} = \frac{\theta_1\overline{Z}_2'\text{-}\left[(P_1+P_3P)\left(\gamma_{20}+\theta_1^2\right)+\{P_2+P_3(1\text{-}P)\}\theta_1\theta_2\right]\overline{Z}_1}{\left[\{P_2+P_3(1\text{-}P)\}\theta_1\gamma_{02}\text{-}(P_1+P_3P)\theta_2\gamma_{20}\right]} \tag{21}$$

I have the following theorems.

Theorem 1: $\hat{\mu}_{Y1}$ is an unbiased estimator of the population mean μ_{Y1}.

$$E(\hat{\mu}_{Y1}) = \mu_{Y1} \tag{22}$$

Proof: Taking expectation on both sides of Eq. (20), I have

$$E(\hat{\mu}_{Y1}) = \frac{\left[(P_1+P_3P)\theta_1\theta_2+\{P_2+P_3(1\text{-}P)\}\left(\gamma_{02}+\theta_2^2\right)\right]E(\overline{Z}_1)\text{-}\theta_2 E\left(\overline{Z}_2'\right)}{\left[\{P_2+P_3(1\text{-}P)\}\theta_1\gamma_{02}\text{-}(P_1+P_3P)\theta_2\gamma_{20}\right]}$$

$$= \frac{\left[(P_1+P_3P)\theta_1\theta_2+\{P_2+P_3(1\text{-}P)\}\left(\gamma_{02}+\theta_2^2\right)\right]1/n\sum_{i=1}^n E(Z_{1i})\text{-}\theta_2 1/n\sum_{i=1}^n E\left(Z_{2i}'\right)}{\left[\{P_2+P_3(1\text{-}P)\}\theta_1\gamma_{02}\text{-}(P_1+P_3P)\theta_2\gamma_{20}\right]}$$

$$= \frac{\begin{matrix}\left[(P_1+P_3P)\theta_1\theta_2+\{P_2+P_3(1\text{-}P)\}\left(\gamma_{02}+\theta_2^2\right)\right]\sum\limits_{i=1}^{n}(\theta_1\mu_{Y1}+\theta_2\mu_{Y2}) \\ \text{-}\ \theta_2\sum_{i=1}^n\left[(P_1+P_3P)\left\{\left(\gamma_{20}+\theta_1^2\right)\mu_{Y1}+\theta_1\theta_2\mu_{Y2}\right\}+\{P_2+P_3(1\text{-}P)\}\left\{\theta_1\theta_2\mu_{Y1}+\left(\gamma_{02}+\theta_2^2\right)\mu_{Y2}\right\}\right]\end{matrix}}{n\left[\{P_2+P_3(1\text{-}P)\}\theta_1\gamma_{02}\text{-}(P_1+P_3P)\theta_2\gamma_{20}\right]}$$

After simplification, I get

$$=\frac{[\{P_2+P_3(1\text{-}P)\}\theta_1\gamma_{02}\text{-}(P_1+P_3P)\theta_2\gamma_{20}]\mu_{Y1}}{[\{P_2+P_3(1\text{-}P)\}\theta_1\gamma_{02}\text{-}(P_1+P_3P)\theta_2\gamma_{20}]}=\mu_{Y1}.$$

which completes the proof.

Theorem 2: $\hat{\mu}_{Y2}$ is an unbiased estimator of the population mean μ_{Y2}.

$$E(\hat{\mu}_{Y2})=\mu_{Y2} \tag{23}$$

Proof: Taking expectation on both sides of Eq. (21), I have

$$\hat{\mu}_{Y2}=\frac{\theta_1 E\left(\overline{Z}_2'\right)\text{-}\left[(P_1+P_3P)\left(\gamma_{20}+\theta_1^2\right)+\{P_2+P_3(1\text{-}P)\}\theta_1\theta_2\right]E\left(\overline{Z}_1\right)}{[\{P_2+P_3(1\text{-}P)\}\theta_1\gamma_{02}\text{-}(P_1+P_3P)\theta_2\gamma_{20}]}$$

Similarly, following the pattern as given in Theorem 1, I obtain

$$=\frac{[\{P_2+P_3(1\text{-}P)\}\theta_1\gamma_{02}\text{-}(P_1+P_3P)\theta_2\gamma_{20}]\mu_{Y2}}{[\{P_2+P_3(1\text{-}P)\}\theta_1\gamma_{02}\text{-}(P_1+P_3P)\theta_2\gamma_{20}]}=\mu_{Y2}.$$

hence, it is proved.

Theorem 3: The variance of the unbiased estimator $\hat{\mu}_{Y1}$ is given by:

$$V(\hat{\mu}_{Y1})=\frac{\begin{array}{l}\left[(P_1+P_3P)\theta_1\theta_2+\{P_2+P_3(1\text{-}P)\}\left(\gamma_{02}+\theta_2^2\right)\right]^2\sigma^2_{Z_1}+\theta_2^2\sigma^2_{Z_2'}\\ \text{-}\ 2\theta_2\left[(P_1+P_3P)\theta_1\theta_2+\{P_2+P_3(1\text{-}P)\}\left(\gamma_{02}+\theta_2^2\right)\right]\sigma_{Z_1Z_2'}\end{array}}{n[\{P_2+P_3(1\text{-}P)\}\theta_1\gamma_{02}\text{-}(P_1+P_3P)\theta_2\gamma_{20}]^2} \tag{24}$$

where $\sigma^2_{Z_1}=\gamma_{20}\left(\sigma^2_{Y1}+\mu^2_{Y1}\right)+\gamma_{02}\left(\sigma^2_{Y2}+\mu^2_{Y2}\right)+\theta_1^2\sigma^2_{Y1}+\theta_2^2\sigma^2_{Y2}+2\theta_1\theta_2\sigma_{Y1}\sigma_{Y2}$,

$$\begin{aligned}\sigma^2_{Z_2'}=&\left(\sigma^2_{Y1}+\mu^2_{Y1}\right)\left[(P_1+P_3P)\left(\gamma_{40}+4\gamma_{30}\theta_1+6\gamma_{20}\theta_1^2+\theta_1^4\right)+\{P_2+P_3(1\text{-}P)\}\left(\gamma_{20}+\theta_1^2\right)\left(\gamma_{02}+\theta_2^2\right)\right]\\&+\left(\sigma^2_{Y2}+\mu^2_{Y2}\right)\left[\{P_2+P_3(1\text{-}P)\}\left(\gamma_{04}+4\gamma_{03}\theta_2+6\gamma_{02}\theta_2^2+\theta_2^4\right)+(P_1+P_3P)\left(\gamma_{20}+\theta_1^2\right)\left(\gamma_{02}+\theta_2^2\right)\right]\\&+2(\sigma_{Y1}\sigma_{Y2}+\mu_{Y1}\mu_{Y2})\left[(P_1+P_3P)\theta_2\left(\gamma_{30}+3\gamma_{20}\theta_1+\theta_1^3\right)+\{P_2+P_3(1\text{-}P)\}\theta_1\left(\gamma_{03}+3\gamma_{02}\theta_2+\theta_2^3\right)\right]\\&\text{-}\ \left[\left\{(P_1+P_3P)\left(\gamma_{20}+\theta_1^2\right)+\{P_2+P_3(1\text{-}P)\}\theta_1\theta_2\right\}\mu_{Y1}\right.\\&\left.+\left\{(P_1+P_3P)\theta_1\theta_2+\{P_2+P_3(1\text{-}P)\}\left(\gamma_{02}+\theta_2^2\right)\right\}\mu_{Y2}\right]^2\end{aligned}$$

and

$$\begin{aligned}\sigma_{Z_1Z_2'}=&\left(\sigma^2_{Y1}+\mu^2_{Y1}\right)\left[(P_1+P_3P)\left(\gamma_{30}+3\gamma_{20}\theta_1+\theta_1^3\right)+\{P_2+P_3(1\text{-}P)\}\theta_2\left(\gamma_{20}+\theta_1^2\right)\right]\\&+\left(\sigma^2_{Y2}+\mu^2_{Y2}\right)\left[(P_1+P_3P)\theta_1\left(\gamma_{02}+\theta_2^2\right)+\{P_2+P_3(1\text{-}P)\}\left(\gamma_{03}+3\gamma_{02}\theta_2+\theta_2^3\right)\right]\\&+2(\sigma_{Y1}\sigma_{Y2}+\mu_{Y1}\mu_{Y2})\left[(P_1+P_3P)\theta_2\left(\gamma_{20}+\theta_1^2\right)+\{P_2+P_3(1\text{-}P)\}\theta_1\left(\gamma_{02}+\theta_2^2\right)\right]\\&\text{-}\ (\theta_1\mu_{Y1}+\theta_2\mu_{Y12})\left[\left\{(P_1+P_3P)\left(\gamma_{20}+\theta_1^2\right)+\{P_2+P_3(1\text{-}P)\}\theta_1\theta_2\right\}\mu_{Y1}\right.\\&\left.+\left\{(P_1+P_3P)\theta_1\theta_2+\{P_2+P_3(1\text{-}P)\}\left(\gamma_{02}+\theta_2^2\right)\right\}\mu_{Y2}\right]\end{aligned}$$

The proof is given in Appendix.

Theorem 4: The variance of the unbiased estimator $\hat{\mu}_{Y2}$ is given by:

$$V(\hat{\mu}_{Y2})=\frac{\begin{array}{l}\theta_1^2\sigma^2_{Z_2'}+\left[(P_1+P_3P)\left(\gamma_{20}+\theta_1^2\right)+\{P_2+P_3(1\text{-}P)\}\theta_1\theta_2\right]^2\sigma^2_{Z_1}\\ \text{-}\ 2\theta_1\left[(P_1+P_3P)\left(\gamma_{20}+\theta_1^2\right)+\{P_2+P_3(1\text{-}P)\}\theta_1\theta_2\right]\sigma_{Z_1Z_2}\end{array}}{n[\{P_2+P_3(1\text{-}P)\}\theta_1\gamma_{02}\text{-}(P_1+P_3P)\theta_2\gamma_{20}]^2} \tag{25}$$

Proof: The proof is similar as given in Theorem 3.

In the next section, I discuss a privacy protection measure to compare the respondent's privacy protection and efficiency for the considered model.

3. Privacy protection measure

A number of measures have been introduced in the literature to estimate the performance of competitive strategies taking into account both efficiency and respondent privacy protection. For a discussion on privacy protection measures for randomized response survey of stigmatizing character, see Lanke [15], Leyseiffer and Warner [16], Bhargava and Singh [17] and among other. These measures of privacy protection are based on the qualitative characters.

When dealing with quantitative sensitive variable, the respondent privacy is conserved by asking interviewees to algebraically scramble the true response by means of a coding mechanism. Respondent's privacy protection measures for quantitative sensitive variable have been investigated by Diana and perri [18] and Zhimin et al. [19] which is based on the square of correlation coefficient i.e. $\rho^2_{y\theta} \in [0,\ 1]$. Later, Diana and Perri [20] introduced the new measure of privacy protection of respondents by using auxiliary variable. These measures are normalized with zero (one) denoting maximum (minimum) privacy protection. Recently, Singh et al. [11] considered the case when no auxiliary variable is available in the procedure and studied the normalized measure of respondent privacy. This normalized measure allows researchers to attain a trade-off between efficiency and privacy. Moreover, it is worth remarking that if one procedure is more efficient than other, then it will be less protective. Thus, all the provided measures using the randomized procedure for the privacy protection, they have concluded for a measure of respondent's privacy protection having a trade-off between these two aspects.

$$\tau = 1\text{-}\rho^2_{y\theta} \tag{26}$$

The values of τ closer to 1 indicates more privacy protection and greater cooperation may be expected using randomized response models while τ closer to zero denotes that the privacy protection is completely violated. Now, I use this normalized measure for comparing the trade-off between efficiency and privacy protection.

In the proposed model, there are two quantitative sensitive variables Y_{1i} and Y_{2i} associated with the second observed response Z'_{2i}. Following Section 2, I compute the square of correlation coefficients between the second observed response Z'_{2i} and quantitative sensitive variables Y_{1i} and Y_{2i} respectively, and are given as:

$$\rho^2_{y_{1i}Z'_{2i}} = \frac{\left[\left(P_1 + P_3P\right)\left\{\left(\gamma_{20} + \theta_1^2\right)\sigma^2_{Y1} + \theta_1\theta_2\sigma_{Y1}\sigma_{Y2}\right\} + \left\{P_2 + P_3(1\text{-}P)\right\}\left\{\theta_1\theta_2\sigma^2_{Y1} + \left(\gamma_{20} + \theta_2^2\right)\sigma_{Y1}\sigma_{Y2}\right\}\right]^2}{\sigma^2_{Y1}\sigma^2_{Z'_2}} \tag{27}$$

and

$$\rho^2_{y_{2i}Z'_{2i}} = \frac{\left[\left(P_1 + P_3P\right)\left\{\left(\gamma_{20} + \theta_1^2\right)\sigma_{Y1}\sigma_{Y2} + \theta_1\theta_2\sigma^2_{Y2}\right\} + \left\{P_2 + P_3(1\text{-}P)\right\}\left\{\theta_1\theta_2\sigma_{Y1}\sigma_{Y2} + \left(\gamma_{20} + \theta_2^2\right)\sigma^2_{Y2}\right\}\right]^2}{\sigma^2_{Y2}\sigma^2_{Z'_2}} \tag{28}$$

where $\sigma^2_{Z'_2}$ is given in Theorem 3.

Now, I define the measure of respondent's privacy protection associated with the proposed second response Z'_{2i} as:

$$\tau_{PJ} = 1\text{-}\rho^2_{y_{Ji}Z'_{2i}}, J = 1,\ 2 \tag{29}$$

I also define the square of correlation coefficients for the Ahmed et al. [14] response model. In the case of Ahmed et al. [14] model, there are also two quantitative sensitive variables associated with the second observed response. Thus, the square of correlation coefficients between the second observed response and quantitative sensitive variables Y_{1i} and Y_{2i} are respectively given by:

$$\rho^2_{Y_{1i}Z_{2i}} = \frac{\left[P\left\{\left(\gamma_{20}+\theta_1^2\right)\sigma^2_{Y1}+\theta_1\theta_2\sigma_{Y1}\sigma_{Y2}\right\}+(1\text{-}P)\left\{\theta_1\theta_2\sigma^2_{Y1}+\left(\gamma_{20}+\theta_2^2\right)\sigma_{Y1}\sigma_{Y2}\right\}\right]^2}{\sigma^2_{Y1}\sigma^2_{Z_2}} \tag{30}$$

$$\rho^2_{Y_{2i}Z_{2i}} = \frac{\left[P\left\{\left(\gamma_{20}+\theta_1^2\right)\sigma_{Y1}\sigma_{Y2}+\theta_1\theta_2\sigma^2_{Y2}\right\}+(1\text{-}P)\left\{\theta_1\theta_2\sigma_{Y1}\sigma_{Y2}+\left(\gamma_{20}+\theta_2^2\right)\sigma^2_{Y2}\right\}\right]^2}{\sigma^2_{Y2}\sigma^2_{Z_2}} \tag{31}$$

where

$$\begin{aligned}\sigma^2_{Z_2} = {} & \left(\sigma^2_{Y1}+\mu^2_{Y1}\right)\left[P\left(\gamma_{40}+4\gamma_{30}\theta_1+6\gamma_{20}\theta_1^2+\theta_1^4\right)+(1\text{-}P)\left(\gamma_{20}+\theta_1^2\right)\left(\gamma_{02}+\theta_2^2\right)\right] \\ & +\left(\sigma^2_{Y2}+\mu^2_{Y2}\right)\left[(1\text{-}P)\left(\gamma_{04}+4\gamma_{03}\theta_2+6\gamma_{02}\theta_2^2+\theta_2^4\right)+P\left(\gamma_{20}+\theta_1^2\right)\left(\gamma_{02}+\theta_2^2\right)\right] \\ & +2(\sigma_{Y1}\sigma_{Y2}+\mu_{Y1}\mu_{Y2})\left[P\theta_2\left(\gamma_{30}+3\gamma_{20}\theta_1+\theta_1^3\right)+(1\text{-}P)\theta_1\left(\gamma_{03}+3\gamma_{02}\theta_2+\theta_2^3\right)\right] \\ & \text{-}\left[\left\{P\left(\gamma_{20}+\theta_1^2\right)+(1\text{-}P)\theta_1\theta_2\right\}\mu_{Y1}+\left\{P\theta_1\theta_2+(1\text{-}P)\left(\gamma_{02}+\theta_2^2\right)\right\}\mu_{Y2}\right]^2\end{aligned}$$

I also define the measure of respondent's privacy protection for Ahmed et al. [14] as:

$$\tau_{AJ} = 1\text{-}\rho^2_{y_{Ji}Z_{2i}}, J = 1,\ 2 \tag{32}$$

In the next section, I investigate the performance of the proposed model with respect to Ahmed et al. [14] model in terms of relative efficiency and privacy protection under different parametric situations.

4. Efficiency vs privacy protection

The relative efficiency (RE) of the proposed estimators $\hat{\mu}_{Y1}$ and $\hat{\mu}_{Y2}$ over Ahmed et al. [14] the estimators $\hat{\mu}_{AY1}$ and $\hat{\mu}_{AY2}$ are respectively given by:

$$RE_J\left(\hat{\mu}_{YJ},\ \hat{\mu}_{AYJ}\right) = \frac{V\left(\hat{\mu}_{AYJ}\right)}{V\left(\hat{\mu}_{YJ}\right)}, J = 1,\ 2 \tag{33}$$

To have a possible trade-off between relative efficiency and privacy protection of respondents, I consider the parametric values in this manner that the relative efficiencies are maximum and expect greater privacy protection of respondents. I decided to take $P = 0.6$, $\mu_{Y1} = 25–45$ with a step 5 and $\mu_{Y2} = 35–55$ with a step 5, five values of θ_1 and θ_2, equal to 2–10 and 4–16 with a increment 2 and 3 respectively, $\sigma_{Y1} = 7$, $\sigma_{Y2} = 5$, $\gamma_{20} = 2$, $\gamma_{02} = 9$, $\gamma_{30} = 1.5$, $\gamma_{03} = 1.2$, $\gamma_{40} = 3.2$ and $\gamma_{04} = 3.5$. I have also chosen different values of probabilities $P_i \ (i = 1, \ 2, \ 3)$ and presented in **Tables 1** and **2**.

Tables 1 and **2** show how the proposed model works in term of efficiency along with privacy protection. For the situation under investigate it emerges that the proposed model based on blank card method is more efficient than Ahmed et al. [14] model. Hence, the finding results, which are worth discussing, are described in the following points.

1. It is observed from **Tables 1** and **2** that the proposed estimators $\hat{\mu}_{YJ} \ (J = 1, \ 2)$ almost equally efficient in term of efficiencies and privacy protection of respondents.

2. From **Tables 1** and 2, it may be observed that the values of relative efficiencies of the proposed estimators $\hat{\mu}_{YJ} \ (J = 1, \ 2)$ with respect to Ahmed et al. [14] estimators $\hat{\mu}_{AYJ} \ (J = 1, \ 2)$ are more than 1 for all case.

3. The behaviour of the estimators in **Tables 1** and **2**, indicates that the highest efficiency 2.03 attains when $P_i = 0.20 \ (i = 1, \ 2)$ and $P_i = 0.60$ with corresponding values of $\theta_1 = 2$ and $\theta_2 = 4$ while the minimum efficiency 1.13

P_1	P_2	P_3	θ_1	θ_2	μ_{Y1}	μ_{Y2}	RE_1	τ_{P1}	τ_{A1}
0.20	0.20	0.60	2	4	25	35	2.03	0.9762	0.9765
			4	7	30	40	1.72	0.9572	0.9570
			6	10	35	45	1.65	0.9399	0.9395
			8	13	40	50	1.62	0.9277	0.9272
			10	16	45	55	1.60	0.9206	0.9201
0.15	0.15	0.70	2	4	25	35	1.74	0.9763	0.9765
			4	7	30	40	1.52	0.9572	0.9570
			6	10	35	45	1.47	0.9398	0.9395
			8	13	40	50	1.45	0.9276	0.9272
			10	16	45	55	1.44	0.9205	0.9201
0.10	0.10	0.80	2	4	25	35	1.47	0.9764	0.9765
			4	7	30	40	1.33	0.9571	0.9570
			6	10	35	45	1.30	0.9397	0.9395
			8	13	40	50	1.29	0.9275	0.9272
			10	16	45	55	1.28	0.9204	0.9201
0.05	0.05	0.90	2	4	25	35	1.22	0.9764	0.9765
			4	7	30	40	1.16	0.9571	0.9570
			6	10	35	45	1.15	0.9396	0.9395
			8	13	40	50	1.14	0.9274	0.9272
			10	16	45	55	1.13	0.9203	0.9201

Table 1.
Relative efficiency of the proposed estimator $\hat{\mu}_{Y_1}$ with respect to Ahmed et al. [14] estimator $\hat{\mu}_{AY_1}$ and privacy protection of the τ_{P_1} and τ_{A_1}.

attains when $P_i = 0.05$ $(i = 1, 2)$ and $P_i = 0.90$ with corresponding values of $\theta_1 = 10$ and $\theta_2 = 16$.

4. It is also seen that the values of relative efficiencies are decreasing as the values of θ_J and μ_{YJ} $(J = 1, 2)$ increase for the fix values of P_i $(i = 1, 2, 3)$.

5. However, it is observed that for the fix values of θ_1, θ_2, μ_{Y1} and μ_{Y2} the values of RE_i $(i = 1, 2)$ are decreasing as the values of P_i $(i = 1, 2)$ decrease.

6. Furthermore, it can be interpreted that with the increase in the values of P_i $(i = 1, 2)$ while decrease in the values of P_3, θ_1, θ_2, μ_{Y1} and μ_{Y2} there is a increasing pattern in the values of RE_i $(i = 1, 2)$.

7. From **Tables 1** and **2**, it is clear that the measure of privacy protection of proposed randomized response model and Ahmed et al. [14] model is closer to one for all the cases, which indicate maximum privacy protection of respondents.

8. It is further observed that the degree of privacy protection of proposed randomized response model and Ahmed et al. [14] model decreasing with the values of θ_J and μ_{YJ} $(J = 1, 2)$ increase.

9. From **Tables 1** and **2**, it may also be seen that the value of respondent's privacy protection are showing an increasing trend with the increase in the values of P_i $(i = 1, 2)$ while decrease in the values of P_3, θ_1, θ_2, μ_{Y1} and μ_{Y2}.

10. However, it is visible that the proposed model is more efficient than Ahmed et al. [14] model but less protective only when $\theta_1 = 2$ and $\theta_2 = 4$. The model which provides more efficiency yields less privacy protection. Hence, I conclude that

$$V(\hat{\mu}_{AYJ}) > V(\hat{\mu}_{YJ}), J = 1, 2$$

and

$$\tau_{AJ} > \tau_{PJ}, J = 1, 2$$

Hence, I conclude that small difference in efficiency may procure substantial improvement in privacy protection of respondent. Thus, our comparisons underline the good performance, in terms of efficiency and respondent's privacy protection.

11. Therefore, the proposed randomized response model under the blank card method may be declared to be best for estimating the mean of two quantitative sensitive variables and thus may be recommended to the survey practitioners whenever they deal with extremely sensitive characteristics.

To judge the performance of the proposed model, I consider a real data CO124 of $N = 124$ units of Sarndal et al. [21]. A random sample of size $n = 30$ units are drawn from the CO124 population. Let Y_1, Y_2 and X be the import, export and military expenditure in the state of U.S. during the year 1983, 1983 and 1981 respectively. The parametric ranges of quantitative sensitive variables Y_1 and Y_2 and non-sensitive variable X have been found by using t-test and chi–square test, which are terms as $\mu_{Y1} \in (331.60, 567.66)$, $\mu_{Y2} \in (242.56, 440.30)$, $\mu_X \in (43, 171.16)$, $\sigma_{Y1} \in (256.03, 432.17)$, $\sigma_{Y2} \in (214.47, 362.02)$ and $\sigma_X \in (138.90, 234.46)$.

P_1	P_2	P_3	θ_1	θ_2	μ_{Y1}	μ_{Y2}	RE_2	τ_{P2}	τ_{A2}
0.20	0.20	0.60	2	4	25	35	1.98	0.9762	0.9765
			4	7	30	40	1.70	0.9572	0.9570
			6	10	35	45	1.64	0.9399	0.9395
			8	13	40	50	1.62	0.9277	0.9272
			10	16	45	55	1.60	0.9206	0.9201
0.15	0.15	0.70	2	4	25	35	1.71	0.9763	0.9765
			4	7	30	40	1.51	0.9572	0.9570
			6	10	35	45	1.47	0.9398	0.9395
			8	13	40	50	1.45	0.9276	0.9272
			10	16	45	55	1.44	0.9205	0.9201
0.10	0.10	0.80	2	4	25	35	1.45	0.9764	0.9765
			4	7	30	40	1.33	0.9571	0.9570
			6	10	35	45	1.30	0.9397	0.9395
			8	13	40	50	1.29	0.9275	0.9272
			10	16	45	55	1.28	0.9204	0.9201
0.05	0.05	0.90	2	4	25	35	1.21	0.9764	0.9765
			4	7	30	40	1.16	0.9571	0.9570
			6	10	35	45	1.15	0.9396	0.9395
			8	13	40	50	1.14	0.9274	0.9272
			10	16	45	55	1.13	0.9203	0.9201

Table 2.
Relative efficiency of the proposed estimator $\hat{\mu}_{Y_2}$ with respect to Ahmed et al. [14] estimator $\hat{\mu}_{AY_2}$ and privacy protection of the τ_{P2} and τ_{A2}.

The relative efficiencies have been computed for these parameters combinations and presented in **Tables 3–7**.

The behaviour of the estimators in **Tables 3–7** indicate that the proposed estimators perform better than Perri [5] and Ahmed et al. [14] estimators in terms of efficiency.

1. When the proposed estimators $\hat{\mu}_{Y1}$ and $\hat{\mu}_{Y2}$ are compared with Perri [5] estimator $\hat{\mu}_P$, the estimator $\hat{\mu}_{Y2}$ gives lesser efficiency than the estimator $\hat{\mu}_{Y1}$. Also, it is clear from **Tables 5** and **6** that when the proposed estimators $\hat{\mu}_{Y1}$ and $\hat{\mu}_{Y2}$ are compared with Ahmed et al. [14] estimators $\hat{\mu}_{AY1}$ and $\hat{\mu}_{AY2}$, the estimator $\hat{\mu}_{Y1}$ gives lesser efficiency than the estimator $\hat{\mu}_{Y2}$.

2. From the simulation results in **Tables 5** and **6**, it can be interpreted that the values of relative efficiencies are coming out to be near 1 when $P = 0.1$, this is the cost to be paid for perturbing the data, so that privacy of respondents is protected.

3. Further, it is observed that for the fix values of P_i, θ_i, μ_{Y_i} and σ_{Yi} $(i = 1, 2, 3)$ the value of relative efficiencies are decreasing in **Tables 3** and **4** while in **Tables 5** and **6** it is increasing as the values of P increase.

The rest of the results can be read out from the given tables.

P_1	P_2	P_3	θ_1	θ_2	σ_{Y1}	σ_{Y2}	μ_{Y1}	μ_{Y2}	P				
									0.1	**0.2**	**0.3**	**0.4**	**0.5**
0.20	0.7	0.10	4	1	260	220	340	250	5.90	5.48	5.10	4.77	4.46
							390	300	5.87	5.45	5.08	4.75	4.45
							440	350	5.86	5.45	5.08	4.76	4.46
							490	400	5.86	5.46	5.10	4.78	4.49
							540	440	5.94	5.54	5.18	4.86	4.56
					310	270	340	250	5.29	4.93	4.61	4.32	4.06
							390	300	5.31	4.95	4.63	4.34	4.08
							440	350	5.35	4.99	4.67	4.38	4.12
							490	400	5.40	5.04	4.72	4.43	4.17
							540	440	5.51	5.14	4.82	4.53	4.26
					360	320	340	250	4.84	4.53	4.24	3.99	3.76
							390	300	4.89	4.57	4.29	4.03	3.80
							440	350	4.95	4.63	4.35	4.09	3.86
							490	400	5.02	4.70	4.41	4.15	3.92
							540	440	5.14	4.81	4.52	4.25	4.01
0.15	0.75	0.10	4	1	260	220	340	250	9.17	8.31	7.57	6.94	6.39
							390	300	9.03	8.20	7.49	6.87	6.34
							440	350	8.93	8.13	7.44	6.84	6.32
							490	400	8.87	8.09	7.41	6.83	6.31
							540	440	8.92	8.15	7.48	6.90	6.39
					310	270	340	250	8.04	7.32	6.70	6.16	5.70
							390	300	8.02	7.31	6.70	6.18	5.72
							440	350	8.03	7.33	6.73	6.21	5.75
							490	400	8.06	7.37	6.77	6.25	5.80
							540	440	8.17	7.48	6.89	6.37	5.91
					360	320	340	250	7.23	6.60	6.07	5.60	5.20
							390	300	7.26	6.65	6.11	5.65	5.25
							440	350	7.32	6.71	6.18	5.71	5.31
							490	400	7.40	6.78	6.25	5.79	5.38
							540	440	7.54	6.92	6.38	5.92	5.50
0.10	0.80	0.10	4	1	260	220	340	250	17.10	14.77	12.92	11.42	10.20
							390	300	16.59	14.38	12.62	11.19	10.02
							440	350	16.16	14.06	12.38	11.01	9.88
							490	400	15.81	13.80	12.19	10.88	9.79
							540	440	15.69	13.75	12.18	10.90	9.83
					310	270	340	250	14.60	12.68	11.16	9.92	8.90
							390	300	14.40	12.54	11.06	9.86	8.86
							440	350	14.24	12.44	11.00	9.83	8.85

P_1	P_2	P_3	θ_1	θ_2	σ_{Y1}	σ_{Y2}	μ_{Y1}	μ_{Y2}	P				
									0.1	0.2	0.3	0.4	0.5
							490	400	14.12	12.38	10.97	9.82	8.86
							540	440	14.16	12.45	11.06	9.92	8.97
					360	320	340	250	12.80	11.18	9.89	8.83	7.96
							390	300	12.75	11.17	9.89	8.85	7.99
							440	350	12.74	11.18	9.93	8.90	8.04
							490	400	12.75	11.22	9.98	8.96	8.11
							540	440	12.88	11.36	10.13	9.11	8.26
0.05	0.85	0.10	4	1	260	220	340	250	49.10	37.34	29.54	24.09	20.10
							390	300	46.49	35.56	28.29	23.17	19.43
							440	350	44.15	33.98	27.18	22.38	18.85
							490	400	42.09	32.61	26.23	21.71	18.37
							540	440	40.76	31.78	25.71	21.38	18.16
					310	270	340	250	40.46	31.02	24.73	20.30	17.06
							390	300	39.09	30.12	24.14	19.89	16.77
							440	350	37.84	29.32	23.59	19.53	16.53
							490	400	36.70	28.60	23.13	19.23	16.34
							540	440	36.03	28.23	22.94	19.16	16.34
					360	320	340	250	34.24	26.47	21.26	17.58	14.86
							390	300	34.55	26.05	21.00	17.42	14.77
							440	350	32.92	25.67	20.79	17.30	14.72
							490	400	32.34	25.35	20.61	17.22	14.69
							540	440	32.07	25.27	20.63	17.30	14.81
0.10	0.85	0.05	4	1	260	220	340	250	18.50	17.10	15.86	14.77	13.79
							390	300	17.92	16.59	15.42	14.38	13.45
							440	350	17.42	16.16	15.05	14.06	13.18
							490	400	17.01	15.81	14.75	13.80	12.96
							540	440	16.85	15.69	14.67	13.75	12.93
					310	270	340	250	15.75	14.60	13.58	12.68	11.88
							390	300	15.51	14.40	13.42	12.54	11.76
							440	350	15.31	14.24	13.29	12.44	11.69
							490	400	15.16	14.12	13.20	12.38	11.64
							540	440	15.18	14.16	13.26	12.45	11.72
					360	320	340	250	13.76	12.80	11.94	11.18	10.50
							390	300	13.70	12.75	11.91	11.17	10.50
							440	350	13.67	12.74	11.92	11.18	10.52
							490	400	13.66	12.75	11.94	11.22	10.57
							540	440	13.79	12.88	12.08	11.36	10.71

Table 3.
Relative efficiency of the proposed estimator $\hat{\mu}_{Y_1}$ with respect to Perri [5] estimator $\hat{\mu}_P$ when $\theta_1 = 4$ and $\theta_2 = 1$.

P_1	P_2	P_3	θ_1	θ_2	σ_{Y1}	σ_{Y2}	μ_{Y1}	μ_{Y2}	P				
									0.1	**0.2**	**0.3**	**0.4**	**0.5**
0.20	0.70	0.10	4	1	260	220	340	250	1.58	1.44	1.31	1.20	1.10
							390	300	1.49	1.36	1.24	1.14	1.05
							440	350	1.43	1.31	1.20	1.10	1.01
							490	400	1.39	1.27	1.16	1.07	0.98
							540	440	1.35	1.23	1.13	1.04	0.96
					310	270	340	250	1.56	1.42	1.31	1.20	1.11
							390	300	1.49	1.36	1.25	1.15	1.06
							440	350	1.43	1.31	1.20	1.11	1.02
							490	400	1.40	1.28	1.17	1.08	1.00
							540	440	1.36	1.24	1.14	1.05	0.97
					360	320	340	250	1.54	1.41	1.30	1.20	1.11
							390	300	1.48	1.36	1.25	1.15	1.07
							440	350	1.43	1.32	1.21	1.12	1.04
							490	400	1.40	1.29	1.18	1.09	1.01
							540	440	1.36	1.25	1.15	1.07	0.99
0.15	0.75	0.10	4	1	260	220	340	250	2.69	2.39	2.14	1.93	1.74
							390	300	2.52	2.25	2.02	1.82	1.64
							440	350	2.40	2.14	1.93	1.74	1.57
							490	400	2.31	2.07	1.86	1.68	1.53
							540	440	2.22	2.00	1.83	1.63	1.48
					310	270	340	250	2.59	2.32	2.08	1.88	1.71
							390	300	2.46	2.20	1.98	1.79	1.63
							440	350	2.36	2.11	1.91	1.73	1.57
							490	400	2.28	2.05	1.85	1.68	1.53
							540	440	2.21	1.99	1.80	1.63	1.48
					360	320	340	250	2.51	2.25	2.04	1.85	1.68
							390	300	2.40	2.16	1.95	1.77	1.62
							440	350	2.32	2.09	1.89	1.72	1.57
							490	400	2.26	2.03	1.84	1.67	1.53
							540	440	2.19	1.98	1.79	1.63	1.49
0.10	0.80	0.10	4	1	260	220	340	250	5.46	4.64	3.99	3.47	3.04
							390	300	5.05	4.30	3.71	3.24	2.85
							440	350	4.73	4.05	3.50	3.06	2.70
							490	400	4.49	3.85	3.35	2.93	2.59
							540	440	4.27	3.68	3.20	2.82	2.49
					310	270	340	250	5.11	4.37	3.78	3.31	2.92
							390	300	4.79	4.11	3.57	3.12	2.76
							440	350	4.55	3.91	3.40	2.99	2.64
							490	400	4.35	3.75	3.27	2.88	2.55

P_1	P_2	P_3	θ_1	θ_2	σ_{Y1}	σ_{Y2}	μ_{Y1}	μ_{Y2}	P				
									0.1	0.2	0.3	0.4	0.5
							540	440	4.17	3.60	3.15	2.78	2.47
					360	320	340	250	4.81	4.14	3.60	3.17	2.81
							390	300	4.57	3.94	3.44	3.03	2.69
							440	350	4.38	3.78	3.30	2.91	2.59
							490	400	4.22	3.66	3.20	2.83	2.52
							540	440	4.07	3.53	3.10	2.74	2.44
0.05	0.85	0.10	4	1	260	220	340	250	16.98	12.71	9.90	7.95	6.53
							390	300	15.32	11.54	9.04	7.28	6.01
							440	350	14.00	10.61	8.36	6.77	5.61
							490	400	12.95	9.88	7.82	6.37	5.30
							540	440	12.03	9.23	7.35	6.01	5.03
					310	270	340	250	15.28	11.54	9.06	7.33	6.06
							390	300	14.06	10.67	8.42	6.83	5.67
							440	350	13.06	9.97	7.90	6.44	5.36
							490	400	12.23	9.39	7.48	6.12	5.12
							540	440	11.49	8.86	7.09	5.83	4.89
					360	320	340	250	13.85	10.56	8.36	6.81	5.67
							390	300	12.96	9.92	7.88	6.44	5.38
							440	350	12.20	9.38	7.48	6.13	5.14
							490	400	11.56	8.93	7.15	5.88	4.94
							540	440	10.96	8.50	6.84	5.64	4.76
0.10	0.85	0.05	4	1	260	220	340	250	5.96	5.46	5.02	4.64	4.30
							390	300	5.50	5.05	4.65	4.30	3.99
							440	350	5.14	4.73	4.37	4.05	3.76
							490	400	4.87	4.49	4.15	3.85	3.59
							540	440	4.62	4.27	3.96	3.68	3.43
					310	270	340	250	5.55	5.11	4.71	4.37	4.06
							390	300	5.20	4.79	4.43	4.11	3.82
							440	350	4.93	4.55	4.21	3.91	3.64
							490	400	4.71	4.35	4.03	3.75	3.50
							540	440	4.51	4.17	3.87	3.60	3.37
					360	320	340	250	5.21	4.81	4.45	4.14	3.86
							390	300	4.95	4.57	4.24	3.94	3.67
							440	350	4.73	4.38	4.06	3.78	3.53
							490	400	4.56	4.22	3.92	3.66	3.42
							540	440	4.39	4.07	3.79	3.53	3.31

Table 4.
Relative efficiency of the proposed estimator $\hat{\mu}_{Y_2}$ with respect to Perri [5] estimator $\hat{\mu}_P$ when $\theta_1 = 4$ and $\theta_2 = 1$.

P_1	P_2	P_3	θ_1	θ_2	σ_{Y1}	σ_{Y2}	μ_{Y1}	μ_{Y2}	P				
									0.1	0.3	0.5	0.7	0.9
0.20	0.7	0.10	4	1	260	220	340	250	0.98	1.01	1.11	1.83	1589.20
							390	300	0.98	1.02	1.12	1.89	1728.00
							440	350	0.98	1.02	1.13	1.95	1854.80
							490	400	0.98	1.02	1.14	2.00	1968.00
							540	440	0.98	1.02	1.15	2.05	2064.70
					310	270	340	250	0.98	1.01	1.10	1.74	1446.10
							390	300	0.98	1.01	1.11	1.80	1568.00
							440	350	0.98	1.01	1.12	1.86	1684.80
							490	400	0.98	1.01	1.13	1.91	1793.90
							540	440	0.98	1.01	1.13	1.96	1890.00
					360	320	340	250	0.99	1.01	1.09	1.68	1343.80
							390	300	0.99	1.01	1.10	1.73	1448.40
							440	350	0.99	1.01	1.11	1.78	1552.40
							490	400	0.99	1.01	1.12	1.83	1653.10
							540	440	0.99	1.01	1.12	1.88	1744.10
0.15	0.75	0.10	4	1	260	220	340	250	0.99	1.02	1.12	1.84	1605.30
							390	300	0.99	1.02	1.13	1.91	1746.40
							440	350	0.99	1.02	1.14	1.97	1875.50
							490	400	0.99	1.02	1.15	2.02	1990.80
							540	440	0.99	1.02	1.16	2.07	2089.90
					310	270	340	250	0.99	1.01	1.11	1.76	1459.00
							390	300	0.99	1.01	1.12	1.82	1582.7
							440	350	0.99	1.02	1.12	1.87	1701.50
							490	400	0.99	1.02	1.13	1.92	1812.50
							540	440	0.99	1.02	1.14	1.98	1910.80
					360	320	340	250	0.99	1.01	1.10	1.70	1345.50
							390	300	0.99	1.01	1.11	1.75	1460.70
							440	350	0.99	1.10	1.12	1.80	1566.40
							490	400	0.99	1.01	1.12	1.85	1668.60
							540	440	0.99	1.02	1.13	1.90	1761.50
0.10	0.80	0.10	4	1	260	220	340	250	0.99	1.02	1.13	1.86	1617.30
							390	300	0.99	1.02	1.14	1.92	1760.20
							440	350	0.99	1.02	1.15	1.98	1890.80
							490	400	0.99	1.03	1.15	2.04	2007.60
							540	440	0.99	1.03	1.16	2.09	2108.50
					310	270	340	250	0.99	1.02	1.12	1.77	1468.60
							390	300	0.99	1.02	1.12	1.83	1593.70
							440	350	0.99	1.02	1.13	1.88	1713.90
							490	400	0.99	1.02	1.14	1.94	1826.20

P_1	P_2	P_3	θ_1	θ_2	σ_{Y1}	σ_{Y2}	μ_{Y1}	μ_{Y2}	P				
									0.1	0.3	0.5	0.7	0.9
							540	440	0.99	1.02	1.15	1.99	1926.10
					360	320	340	250	0.99	1.02	1.10	1.70	1362.50
							390	300	0.99	1.02	1.11	1.76	1469.80
							440	350	0.99	1.02	1.12	1.81	1576.60
							490	400	0.99	1.02	1.13	1.86	1680.10
							540	440	0.99	1.02	1.13	1.91	1774.40
0.05	0.85	0.10	4	1	260	220	340	250	1.00	1.03	1.14	1.87	1626.40
							390	300	1.00	1.03	1.15	1.93	1770.50
							440	350	1.00	1.03	1.15	1.99	1902.30
							490	400	1.00	1.03	1.16	2.05	2020.20
							540	440	1.00	1.03	1.17	2.10	2122.40
					310	270	340	250	1.00	1.02	1.12	1.78	1475.70
							390	300	1.00	1.02	1.13	1.84	1601.90
							440	350	1.00	1.02	1.14	1.89	1723.00
							490	400	1.00	1.03	1.14	1.95	1836.40
							540	440	1.00	1.03	1.15	2.00	1937.50
					360	320	340	250	1.00	1.02	1.11	1.71	1368.40
							390	300	1.00	1.02	1.12	1.76	1476.60
							440	350	1.00	1.02	1.12	1.82	1584.20
							490	400	1.00	1.02	1.13	1.87	1688.50
							540	440	1.00	1.02	1.14	1.92	1783.90
0.10	0.85	0.05	4	1	260	220	340	250	0.99	1.02	1.13	1.86	1625.60
							390	300	0.99	1.02	1.14	1.93	1769.60
							440	350	0.99	1.03	1.15	1.99	1901.30
							490	400	0.99	1.03	1.16	2.04	2019.10
							540	440	0.99	1.03	1.17	2.10	2121.20
					310	270	340	250	0.99	1.02	1.12	1.77	1475.10
							390	300	0.99	1.02	1.13	1.83	1601.20
							440	350	0.99	1.02	1.13	1.89	1722.20
							490	400	0.99	1.02	1.14	1.94	1835.50
							540	440	0.99	1.03	1.15	2.00	1936.50
					360	320	340	250	0.99	1.02	1.11	1.71	1367.90
							390	300	0.99	1.02	1.11	1.76	1476.00
							440	350	0.99	1.02	1.12	1.81	1583.50
							490	400	0.99	1.02	1.13	1.86	1687.80
							540	440	0.99	1.02	1.14	1.91	1783.10

Table 5.
Relative efficiency of the proposed estimator $\hat{\mu}_{Y_1}$ with respect to Ahmed et al. *[14] estimator $\hat{\mu}_{AY_1}$ when $\theta_1 = 4$ and $\theta_2 = 1$.*

P_1	P_2	P_3	θ_1	θ_2	σ_{Y1}	σ_{Y2}	μ_{Y1}	μ_{Y2}	P				
									0.1	0.3	0.5	0.7	0.9
0.20	0.7	0.10	4	1	260	220	340	250	0.81	1.17	2.06	6.14	5899.30
							390	300	0.81	1.17	2.08	6.24	6099.10
							440	350	0.81	1.17	2.09	6.33	6289.70
							490	400	0.81	1.17	2.10	6.42	6466.80
							540	440	0.81	1.17	2.11	6.45	6491.60
					310	270	340	250	0.82	1.16	2.04	6.07	5933.10
							390	300	0.81	1.16	2.05	6.17	6103.60
							440	350	0.81	1.16	2.06	6.26	6272.90
							490	400	0.81	1.17	2.08	6.34	6435.10
							540	440	0.81	1.17	2.09	6.38	6461.80
					360	320	340	250	0.82	1.16	2.01	6.01	5971.40
							390	300	0.82	1.16	2.03	6.10	6115.70
							440	350	0.82	1.16	2.04	6.19	6264.40
							490	400	0.82	1.16	2.06	6.27	6411.20
							540	440	0.81	1.16	2.07	6.31	6438.30
0.15	0.75	0.10	4	1	260	220	340	250	0.89	1.29	2.29	6.85	6603.20
							390	300	0.89	1.29	2.31	6.96	6833.40
							440	350	0.89	1.30	2.32	7.07	7052.40
							490	400	0.89	1.30	2.33	7.17	7255.60
							540	440	0.89	1.30	2.35	7.21	7289.80
					310	270	340	250	0.90	1.28	2.25	6.74	6619.70
							390	300	0.90	1.28	2.27	6.86	6818.50
							440	350	0.90	1.29	2.29	6.97	7015.00
							490	400	0.90	1.29	2.31	7.07	7202.90
							540	440	0.89	1.29	2.32	7.12	7240.80
					360	320	340	250	0.90	1.27	2.22	6.65	6643.30
							390	300	0.90	1.28	2.24	6.76	6813.50
							440	350	0.90	1.28	2.26	6.87	6987.80
							490	400	0.90	1.28	2.28	6.97	7159.30
							540	440	0.90	1.29	2.29	7.03	7198.80
0.10	0.80	0.10	4	1	260	220	340	250	0.98	1.42	2.52	7.56	7317.70
							390	300	0.98	1.42	2.54	7.70	7579.10
							440	350	0.98	1.42	2.56	7.82	7827.10
							490	400	0.98	1.43	2.57	7.93	8056.60
							540	440	0.98	1.43	2.59	7.99	8102.10
					310	270	340	250	0.98	1.40	2.47	7.42	7310.90
							390	300	0.98	1.41	2.50	7.56	7539.20
							440	350	0.98	1.41	2.52	7.68	7763.90
							490	400	0.98	1.41	2.54	7.80	7978.10

P_1	P_2	P_3	θ_1	θ_2	σ_{Y1}	σ_{Y2}	μ_{Y1}	μ_{Y2}	P				
									0.1	0.3	0.5	0.7	0.9
							540	440	0.98	1.42	2.55	7.87	8029.50
					360	320	340	250	0.98	1.39	2.43	7.30	7314.40
							390	300	0.98	1.39	2.45	7.43	7512.00
							440	350	0.98	1.40	2.48	7.56	7713.20
							490	400	0.98	1.40	2.50	7.67	7910.40
							540	440	0.98	1.41	2.52	7.75	7964.70
0.05	0.85	0.10	4	1	260	220	340	250	1.06	1.54	2.75	8.28	8040.70
							390	300	1.06	1.55	2.78	8.44	8333.90
							440	350	1.06	1.55	2.80	8.58	8611.10
							490	400	1.06	1.56	2.82	8.70	8866.90
							540	440	1.06	1.56	2.83	8.77	8925.60
					310	270	340	250	1.06	1.52	2.69	8.10	8004.10
							390	300	1.06	1.53	2.72	8.26	8263.00
							440	350	1.06	1.53	2.74	8.40	8516.70
							490	400	1.06	1.54	2.77	8.53	8757.70
							540	440	1.06	1.55	2.79	8.62	8824.70
					360	320	340	250	1.06	1.50	2.63	7.94	7982.30
							390	300	1.06	1.51	2.66	8.09	8208.60
							440	350	1.06	1.51	2.69	8.24	8437.60
							490	400	1.06	1.52	2.72	8.37	8661.30
							540	440	1.06	1.53	2.74	8.47	8732.70
0.10	0.85	0.05	4	1	260	220	340	250	0.99	1.45	2.64	8.07	7968.00
							390	300	0.99	1.46	2.66	8.21	8258.10
							440	350	0.99	1.46	2.68	8.35	8532.40
							490	400	0.99	1.46	2.69	8.47	8785.50
							540	440	0.99	1.47	2.71	8.53	8842.80
					310	270	340	250	0.99	1.44	2.58	7.89	7934.80
							390	300	0.99	1.44	2.61	8.05	8190.50
							440	350	0.99	1.45	2.63	8.19	8441.30
							490	400	0.99	1.45	2.65	8.31	8679.70
							540	440	0.99	1.46	2.67	8.39	8745.00
					360	320	340	250	0.99	1.42	2.53	7.75	7915.80
							390	300	0.99	1.43	2.56	7.89	8139.10
							440	350	0.99	1.43	2.59	8.03	8365.30
							490	400	0.99	1.44	2.61	8.16	8586.30
							540	440	0.99	1.45	2.63	8.25	8655.90

Table 6.
Relative efficiency of the proposed estimator $\hat{\mu}_{Y_2}$ with respect to Ahmed et al. *[14] estimator $\hat{\mu}_{AY_2}$ when $\theta_1 = 4$ and $\theta_2 = 1$.*

P_1	P_2	P_3	θ_1	θ_2	σ_{Y1}	σ_{Y2}	μ_{Y1}	μ_{Y2}	RE_1	RE_2
0.85	0.10	0.05	4	1	260	220	340	250	3.5688	3.5201
							390	300	3.5701	3.5221
							440	350	3.5712	3.5241
							490	400	3.5722	3.5258
							540	440	3.5723	3.5255
					310	270	340	250	3.5691	3.5227
							390	300	3.5702	3.5241
							440	350	3.5712	3.5255
							490	400	3.5721	3.5269
							540	440	3.5722	3.5265
					360	320	340	250	3.5695	3.5251
							390	300	3.5704	3.5260
							440	350	3.5712	3.5270
							490	400	3.5720	3.5281
							540	440	3.5721	3.5276
0.80	0.15	0.05	4	1	260	220	340	250	116.44	105.89
							390	300	117.42	106.52
							440	350	118.22	107.12
							490	400	118.87	107.66
							540	440	119.17	107.59
					310	270	340	250	115.91	106.64
							390	300	116.84	107.08
							440	350	117.63	107.54
							490	400	118.30	107.94
							540	440	118.65	107.87
					360	320	340	250	115.48	107.35
							390	300	116.34	107.64
							440	350	117.10	107.97
							490	400	117.77	108.31
							540	440	118.15	108.18
0.75	0.20	0.05	4	1	260	220	340	250	355.18	328.81
							390	300	363.11	332.28
							440	350	369.70	335.58
							490	400	375.11	338.62
							540	440	378.32	338.29
					310	270	340	250	348.93	332.66
							390	300	356.59	335.15
							440	350	363.24	337.70
							490	400	368.94	340.18
							540	440	372.59	339.68

P_1	P_2	P_3	θ_1	θ_2	σ_{Y1}	σ_{Y2}	μ_{Y1}	μ_{Y2}	RE_1	RE_2
					360	320	340	250	343.89	336.31
							390	300	351.03	338.03
							440	350	357.49	339.94
							490	400	363.22	341.91
							540	440	367.14	341.23
0.70	0.25	0.10	4	1	260	220	340	250	637.70	649.04
							390	300	661.97	658.47
							440	350	682.56	667.49
							490	400	699.81	675.82
							540	440	711.27	675.11
					310	270	340	250	616.17	658.67
							390	300	639.28	665.62
							440	350	659.84	672.73
							490	400	677.80	679.63
							540	440	690.56	678.47
					360	320	340	250	599.19	667.92
							390	300	620.45	672.86
							440	350	640.14	678.31
							490	400	657.98	683.88
							540	440	671.33	682.30
0.65	0.30	0.05	4	1	260	220	340	250	894.78	1048.40
							390	300	941.37	1067.30
							440	350	981.73	1085.40
							490	400	1016.20	1102.10
							540	440	1040.80	1101.10
					310	270	340	250	851.25	1066.00
							390	300	894.75	1080.20
							440	350	934.31	1094.70
							490	400	969.53	1108.80
							540	440	996.20	1106.90
					360	320	340	250	817.76	1083.00
							390	300	857.09	1093.40
							440	350	894.29	1104.70
							490	400	928.64	1116.30
							540	440	955.85	1113.70
0.60	0.35	0.05	4	1	260	220	340	250	1098.70	1512.50
							390	300	1168.00	1544.30
							440	350	1229.00	1574.90
							490	400	1281.70	1603.20
							540	440	1321.40	1602.10

P_1	P_2	P_3	θ_1	θ_2	σ_{Y1}	σ_{Y2}	μ_{Y1}	μ_{Y2}	RE_1	RE_2
					310	270	340	250	1032.30	1539.30
							390	300	1095.80	1563.70
							440	350	1154.50	1588.50
							490	400	1207.60	1612.70
							540	440	1249.60	1610.30
					360	320	340	250	917.00	1565.30
							390	300	981.16	1583.60
							440	350	1042.50	1603.40
							490	400	1099.60	1623.60
							540	440	1146.60	1620.10
0.55	0.40	0.05	4	1	260	220	340	250	1250.10	2029.90
							390	300	1338.80	2077.90
							440	350	1418.00	2124.00
							490	400	1487.20	2167.00
							540	440	1541.10	2166.20
					310	270	340	250	1163.40	2065.80
							390	300	1243.70	2103.40
							440	350	1318.80	2141.50
							490	400	1387.50	2178.50
							540	440	1443.50	2176.20
					360	320	340	250	1099.00	2101.10
							390	300	1169.70	2129.90
							440	350	1238.50	2160.80
							490	400	1303.50	2192.10
							540	440	1358.60	2188.40
0.50	0.45	0.5	4	1	260	220	340	250	1359.40	2591.10
							390	300	1463.50	2658.30
							440	350	1557.20	2723.00
							490	400	1639.70	2783.20
							540	440	1705.70	2783.60
					310	270	340	250	1256.30	2635.20
							390	300	1349.60	2688.70
							440	350	1437.60	2742.80
							490	400	1518.70	2795.20
							540	440	1586.30	2794.00
					360	320	340	250	1180.70	2678.70
							390	300	1262.20	2720.60
							440	350	1341.90	2765.20
							490	400	1418.10	2810.20
							540	440	1483.80	2807.20

P_1	P_2	P_3	θ_1	θ_2	σ_{Y1}	σ_{Y2}	μ_{Y1}	μ_{Y2}	RE_1	RE_2
0.45	0.50	0.5	4	1	260	220	340	250	1437.80	3188.60
							390	300	1553.50	3277.70
							440	350	1658.20	3363.40
							490	400	1751.00	3443.30
							540	440	1826.70	3445.70
					310	270	340	250	1322.00	3238.70
							390	300	1424.90	3310.70
							440	350	1522.60	3383.30
							490	400	1613.10	3453.60
							540	440	1689.90	3454.50
					360	320	340	250	1237.80	3288.40
							390	300	1327.10	3345.90
							440	350	1415.10	3406.70
							490	400	1499.60	3467.70
							540	440	1573.50	3466.80
0.40	0.55	0.5	4	1	260	220	340	250	1494.20	3816.00
							390	300	1618.50	3929.40
							440	350	1731.40	4038.20
							490	400	1831.80	4139.70
							540	440	1915.00	4145.40
					310	270	340	250	1368.80	3869.00
							390	300	1478.70	3961.90
							440	350	1583.60	4055.30
							490	400	1681.00	4145.50
							540	440	1764.70	4150.00
					360	320	340	250	1278.00	3922.10
							390	300	1373.10	3997.50
							440	350	1467.10	4076.80
							490	400	1557.60	4155.90
							540	440	1637.80	4158.80
0.35	0.60	0.5	4	1	260	220	340	250	1535.20	4468.10
							390	300	1665.70	4607.70
							440	350	1784.70	4741.50
							490	400	1890.70	4866.30
							540	440	1979.50	4876.50
					310	270	340	250	1402.30	4520.00
							390	300	1517.50	4636.00
							440	350	1627.50	4752.20
							490	400	1730.00	4864.20
							540	440	1819.00	4873.80

P_1	P_2	P_3	θ_1	θ_2	σ_{Y1}	σ_{Y2}	μ_{Y1}	μ_{Y2}	RE_1	RE_2
					360	320	340	250	1306.70	4573.00
							390	300	1406.00	4668.60
							440	350	1504.40	4768.30
							490	400	1599.30	4867.60
							540	440	1684.00	4876.00

Table 7.
Relative efficiency of the proposed estimators $\hat{\mu}_{Y_1}$ and $\hat{\mu}_{Y_2}$ with respect to Ahmed et al. [14] estimators $\hat{\mu}_{AY_1}$ and $\hat{\mu}_{AY_2}$ respectively when $\theta_1 = 4$, $\theta_2 = 1$ and $P = 0.9$.

5. Conclusions

The main objective of this paper is to estimate the population means of two quantitative sensitive variables. It is to be pointed out that the proposed model is more efficient in terms of relative efficiencies and respondent's privacy protection. Therefore, these results advocate that the proposed technique is appreciatively favourable in obtaining the truthful response from the respondents.

Appendix

Proof: Given that $E(S_1) = \theta_1$ and $E(S_2) = \theta_2$. Following Singh [9], I define

$$\gamma_{rs} = E[S_1\text{-}\theta_1]^r[S_2\text{-}\theta_2]^s \tag{34}$$

Then due to independence of the scramble variables, I have

$$E(S_1^2) = \gamma_{20} + \theta_1^2 \tag{35}$$

$$E(S_1^3) = \gamma_{30} + 3\gamma_{20}\theta_1 + \theta_1^3 \tag{36}$$

$$E(S_1^4) = \gamma_{40} + 4\gamma_{30}\theta_1 + 6\gamma_{20}\theta_1^2 + \theta_1^4 \tag{37}$$

$$E(S_2^2) = \gamma_{02} + \theta_2^2 \tag{38}$$

$$E(S_2^3) = \gamma_{03} + 3\gamma_{02}\theta_2 + \theta_2^3 \tag{39}$$

$$E(S_2^4) = \gamma_{04} + 4\gamma_{03}\theta_2 + 6\gamma_{02}\theta_2^2 + \theta_2^4 \tag{40}$$

$$E(S_1S_2) = \theta_1\theta_2 \tag{41}$$

$$E(S_1^2S_2^2) = (\gamma_{20} + \theta_1^2)(\gamma_{02} + \theta_2^2) \tag{42}$$

$$E(S_1^3S_2) = (\gamma_{30} + 3\gamma_{20}\theta_1 + \theta_1^3)\theta_2 \tag{43}$$

and

$$E(S_1S_2^3) = \theta_2(\gamma_{03} + 3\gamma_{02}\theta_2 + \theta_2^3) \tag{44}$$

$$V(\hat{\mu}_{Y1}) = \frac{\left[(P_1+P_3P)\theta_1\theta_2+\{P_2+P_3(1\text{-}P)\}\left(\gamma_{02}+\theta_2^2\right)\right]^2\sum_{i=1}^{n}V(Z_{1i})+\theta_2^2\sum_{i=1}^{n}V(Z'_{2i}) \\ \text{-}\ 2\theta_2\left[(P_1+P_3P)\theta_1\theta_2+\{P_2+P_3(1\text{-}P)\}\left(\gamma_{02}+\theta_2^2\right)\right]\sum_{i=1}^{n}\text{cov}\left(Z_{1i}Z'_{2i}\right)}{n^2[\{P_2+P_3(1\text{-}P)\}\theta_1\gamma_{02}\text{-}(P_1+P_3P)\theta_2\gamma_{20}]^2}$$

$$= \frac{\left[(P_1+P_3P)\theta_1\theta_2+\{P_2+P_3(1\text{-}P)\}\left(\gamma_{02}+\theta_2^2\right)\right]^2\sigma_{Z_1}^2+\theta_2^2\sigma_{Z'_2}^2 \\ \text{-}\ 2\theta_2\left[(P_1+P_3P)\theta_1\theta_2+\{P_2+P_3(1\text{-}P)\}\left(\gamma_{02}+\theta_2^2\right)\right]\sigma_{Z_1Z'_2}}{n[\{P_2+P_3(1\text{-}P)\}\theta_1\gamma_{02}\text{-}(P_1+P_3P)\theta_2\gamma_{20}]^2} \quad (45)$$

where, the variance $\sigma_{Z_1}^2$ is given by:

$$\begin{aligned}
\sigma_{Z_1}^2 &= E\left(Z_{1i}^2\right)\text{-}[E(Z_{1i})]^2 \\
&= E(S_1Y_{1i}+S_2Y_{1i})^2\text{-}[E(S_1Y_{1i}+S_2Y_{1i})]^2 \\
&= E\left(S_1^2Y_{1i}^2+S_2^2Y_{2i}^2+2S_1S_2Y_{1i}Y_{2i}\right)\text{-}[\theta_1\mu_{Y1}+\theta_2\mu_{Y2}]^2 \\
&= \left(\gamma_{20}+\theta_1^2\right)\left(\sigma_{Y1}^2+\mu_{Y1}^2\right)+\left(\gamma_{02}+\theta_2^2\right)\left(\sigma_{Y2}^2+\mu_{Y2}^2\right) \\
&\quad +2\theta_1\theta_2(\sigma_{Y1}\sigma_{Y2}+\mu_{Y1}\mu_{Y2})\text{-}[\theta_1\mu_{Y1}+\theta_2\mu_{Y2}]^2 \\
&= \gamma_{20}\left(\sigma_{Y1}^2+\mu_{Y1}^2\right)+\gamma_{02}\left(\sigma_{Y2}^2+\mu_{Y2}^2\right)+\theta_1^2\sigma_{Y1}^2+\theta_2^2\sigma_{Y2}^2+2\theta_1\theta_2\sigma_{Y1}\sigma_{Y2} \quad (46)
\end{aligned}$$

The variance $\sigma_{Z'_2}^2$ is given by:

$$\begin{aligned}
\sigma_{Z'_2}^2 &= E\left(Z'^2_{2i}\right)\text{-}\left[E\left(Z'_{2i}\right)\right]^2 \\
&= (P_1+P_3P)E\left(S_1^2Y_{1i}+S_1S_2Y_{2i}\right)^2+\{P_2+P_3(1\text{-}P)\}E\left(S_1S_2Y_{1i}+S_2^2Y_{2i}\right)^2 \\
&\text{-}\left[(P_1+P_3P)\left\{\left(\gamma_{20}+\theta_1^2\right)\mu_{Y1}+\theta_1\theta_2\mu_{Y2}\right\}+\{P_2+P_3(1\text{-}P)\}\left\{\theta_1\theta_2\mu_{Y1}+\left(\gamma_{02}+\theta_2^2\right)\mu_{Y2}\right\}\right]^2 \\
&= (P_1+P_3P)E\left(S_1^4Y_{1i}^2+S_1^2S_2^2Y_{2i}^2+2S_1^3S_2Y_{1i}Y_{2i}\right) \\
&\quad +\{P_2+P_3(1\text{-}P)\}E\left(S_1^2S_2^2Y_{1i}^2+S_2^4Y_{2i}^2+2S_1S_2^3Y_{1i}Y_{2i}\right) \\
&\quad \text{-}\left[(P_1+P_3P)\left\{\left(\gamma_{20}+\theta_1^2\right)\mu_{Y1}+\theta_1\theta_2\mu_{Y2}\right\}+\{P_2+P_3(1\text{-}P)\}\left\{\theta_1\theta_2\mu_{Y1}+\left(\gamma_{02}+\theta_2^2\right)\mu_{Y2}\right\}\right]^2 \\
&= (P_1+P_3P)\left[\left(\gamma_{40}+4\gamma_{30}\theta_1+6\gamma_{20}\theta_1^2+\theta_1^4\right)\left(\sigma_{Y1}^2+\mu_{Y1}^2\right)+\left(\gamma_{20}+\theta_1^2\right)\left(\gamma_{02}+\theta_2^2\right)\left(\sigma_{Y2}^2+\mu_{Y2}^2\right)\right. \\
&\quad \left.+2\left(\gamma_{30}+3\gamma_{20}\theta_1+\theta_1^3\right)\theta_2(\sigma_{Y1}\sigma_{Y2}+\mu_{Y1}\mu_{Y2})\right]+\{P_2+P_3(1\text{-}P)\}\left[\left(\gamma_{20}+\theta_1^2\right)\left(\gamma_{02}+\theta_2^2\right)\left(\sigma_{Y1}^2+\mu_{Y1}^2\right)\right. \\
&\quad \left.+\left(\gamma_{04}+4\gamma_{03}\theta_2+6\gamma_{02}\theta_2^2+\theta_2^4\right)\left(\sigma_{Y2}^2+\mu_{Y2}^2\right)+2\theta_1\left(\gamma_{03}+3\gamma_{02}\theta_2+\theta_2^3\right)(\sigma_{Y1}\sigma_{Y2}+\mu_{Y1}\mu_{Y2})\right] \\
&\quad \text{-}\left[(P_1+P_3P)\left\{\left(\gamma_{20}+\theta_1^2\right)\mu_{Y1}+\theta_1\theta_2\mu_{Y2}\right\}+\{P_2+P_3(1\text{-}P)\}\left\{\theta_1\theta_2\mu_{Y1}+\left(\gamma_{02}+\theta_2^2\right)\mu_{Y2}\right\}\right]^2
\end{aligned}$$

After simplification, this gives

$$\begin{aligned}
\sigma_{Z'_2}^2 &= \left(\sigma_{Y1}^2+\mu_{Y1}^2\right)\left[(P_1+P_3P)\left(\gamma_{40}+4\gamma_{30}\theta_1+6\gamma_{20}\theta_1^2+\theta_1^4\right)+\{P_2+P_3(1\text{-}P)\}\left(\gamma_{20}+\theta_1^2\right)\left(\gamma_{02}+\theta_2^2\right)\right] \\
&\quad +\left(\sigma_{Y2}^2+\mu_{Y2}^2\right)\left[\{P_2+P_3(1\text{-}P)\}\left(\gamma_{04}+4\gamma_{03}\theta_2+6\gamma_{02}\theta_2^2+\theta_2^4\right)+(P_1+P_3P)\left(\gamma_{20}+\theta_1^2\right)\left(\gamma_{02}+\theta_2^2\right)\right] \\
&\quad +2(\sigma_{Y1}\sigma_{Y2}+\mu_{Y1}\mu_{Y2})\left[(P_1+P_3P)\theta_2\left(\gamma_{30}+3\gamma_{20}\theta_1+\theta_1^3\right)+\{P_2+P_3(1\text{-}P)\}\theta_1\left(\gamma_{03}+3\gamma_{02}\theta_2+\theta_2^3\right)\right] \\
&\quad \text{-}\left[\left\{(P_1+P_3P)\left(\gamma_{20}+\theta_1^2\right)+\{P_2+P_3(1\text{-}P)\}\theta_1\theta_2\right\}\mu_{Y1}\right. \\
&\quad \left.+\left\{(P_1+P_3P)\theta_1\theta_2+\{P_2+P_3(1\text{-}P)\}\left(\gamma_{02}+\theta_2^2\right)\right\}\mu_{Y2}\right]^2
\end{aligned} \quad (47)$$

and the covariance $\sigma_{Z_1Z_2'}$ between Z_{1i} and Z_{2i}' is given by:

$$\sigma_{Z_1Z_2'} = \text{cov}\left(Z_{1i},\ Z_{2i}'\right) = E\left(Z_{1i}Z_{2i}'\right)\text{-}E\left(Z_{1i}\right)E\left(Z_{2i}'\right)$$

$$= \left(P_1 + P_3P\right)E\left\{\left(S_1Y_{1i} + S_2Y_{2i}\right)\left(S_1^2Y_{1i} + S_1S_2Y_{2i}\right)\right\} + \left\{P_2 + P_{3(1\text{-}P)}\right\}E\left\{\left(S_1Y_{1i} + S_2Y_{2i}\right)\left(S_1S_2Y_{1i} + S_2^2Y_{2i}\right)\right\}$$

$$\text{-}\left(\theta_1\mu_{Y1} + \theta_2\mu_{Y2}\right)\left[\left(P_1 + P_3P\right)\left\{\left(\gamma_{20} + \theta_1^2\right)\mu_{Y1} + \theta_1\theta_2\mu_{Y2}\right\} + \left\{P_2 + P_{3(1\text{-}P)}\right\}\left\{\theta_1\theta_2\mu_{Y1} + \left(\gamma_{02} + \theta_2^2\right)\mu_{Y2}\right\}\right]^2$$

$$= \left(P_1 + P_3P\right)E\left(S_1^3Y_{1i}^2 + 2S_1^2S_2Y_{1i}Y_{2i} + S_1S_2^2Y_{2i}^2\right) + \left\{P_2 + P_{3(1\text{-}P)}\right\}E\left(S_1^2S_2Y_{1i}^2 + 2S_1S_2^2Y_{1i}Y_{2i} + S_2^3Y_{2i}^2\right)$$

$$\text{-}\left(\theta_1\mu_{Y1} + \theta_2\mu_{Y2}\right)\left[\left(P_1 + P_3P\right)\left\{\left(\gamma_{20} + \theta_1^2\right)\mu_{Y1} + \theta_1\theta_2\mu_{Y2}\right\} + \left\{P_2 + P_{3(1\text{-}P)}\right\}\left\{\theta_1\theta_2\mu_{Y1} + \left(\gamma_{02} + \theta_2^2\right)\mu_{Y2}\right\}\right]^2$$

$$= \left(P_1 + P_3P\right)\left[\left(\gamma_{30} + 3\gamma_{20}\theta_1 + \theta_1^3\right)\left(\sigma_{Y1}^2 + \mu_{Y1}^2\right) + 2\left(\gamma_{20} + \theta_1^2\right)\theta_2\left(\sigma_{Y1}\sigma_{Y2} + \mu_{Y1}\mu_{Y2}\right) + \theta_1\left(\gamma_{02} + \theta_2^2\right)\left(\sigma_{Y2}^2 + \mu_{Y2}^2\right)\right]$$

$$+\left\{P_2 + P_{3(1\text{-}P)}\right\}\left[\left(\gamma_{20} + \theta_1^2\right)\theta_2\left(\sigma_{Y1}^2 + \mu_{Y1}^2\right) + 2\theta_1\left(\gamma_{02} + \theta_2^2\right)\left(\sigma_{Y1}\sigma_{Y2} + \mu_{Y1}\mu_{Y2}\right) + \left(\gamma_{03} + 3\gamma_{02}\theta_2 + \theta_2^3\right)\left(\sigma_{Y2}^2 + \mu_{Y2}^2\right)\right]$$

$$\text{-}\left(\theta_1\mu_{Y1} + \theta_2\mu_{Y2}\right)\left[\left(P_1 + P_3P\right)\left\{\left(\gamma_{20} + \theta_1^2\right)\mu_{Y1} + \theta_1\theta_2\mu_{Y2}\right\} + \left\{P_2 + P_{3(1\text{-}P)}\right\}\left\{\theta_1\theta_2\mu_{Y1} + \left(\gamma_{02} + \theta_2^2\right)\mu_{Y2}\right\}\right]^2$$

After some algebra, I get

$$\begin{aligned}\sigma_{Z_1Z_2'} = {} & \left(\sigma_{Y1}^2 + \mu_{Y1}^2\right)\left[(P_1 + P_3P)\left(\gamma_{30} + 3\gamma_{20}\theta_1 + \theta_1^3\right) + \{P_2 + P_3(1\text{-}P)\}\theta_2\left(\gamma_{20} + \theta_1^2\right)\right] \\ & +\left(\sigma_{Y2}^2 + \mu_{Y2}^2\right)\left[(P_1 + P_3P)\theta_1\left(\gamma_{02} + \theta_2^2\right) + \{P_2 + P_3(1\text{-}P)\}\left(\gamma_{03} + 3\gamma_{02}\theta_2 + \theta_2^3\right)\right] \\ & +2(\sigma_{Y1}\sigma_{Y2} + \mu_{Y1}\mu_{Y2})\left[(P_1 + P_3P)\theta_2\left(\gamma_{20} + \theta_1^2\right) + \{P_2 + P_3(1\text{-}P)\}\theta_1\left(\gamma_{02} + \theta_2^2\right)\right] \\ & \text{-}(\theta_1\mu_{Y1} + \theta_2\mu_{Y12})\left[\left\{(P_1 + P_3P)\left(\gamma_{20} + \theta_1^2\right) + \{P_2 + P_3(1\text{-}P)\}\theta_1\theta_2\right\}\mu_{Y1}\right. \\ & \left. +\left\{(P_1 + P_3P)\theta_1\theta_2 + \{P_2 + P_3(1\text{-}P)\}\left(\gamma_{02} + \theta_2^2\right)\right\}\mu_{Y2}\right]\end{aligned} \tag{48}$$

Finally, substituting the expressions given in Eqs. (13), (14) and (15) in Eq. (12), I get the variance of the estimator $\hat{\mu}_{Y1}$ as given in Eq. (24).

Author details

Amod Kumar
Department of Mathematics and Statistics, Swami Vivekanad Subharti University, Meerut, India

*Address all correspondence to: amod.ism01@gmail.com

References

[1] Warner SL. Randomized Response: A survey technique for eliminating evasive answer bias. Journal of the American Statistical Association. 1965; **60**(309):63-69

[2] Greenberg BG, Abul-Ela A, Simmons WR, Horvitz DG. The unrelated question randomized response model: Theoretical Framework. Journal of the American Statistical Association. 1969;**64**:520-539

[3] Greenberg BG, Kuebler RR Jr, Abernathy JR, Horvitz DG. Application of the randomized response technique in obtaining quantitative data. Journal of the American Statistical Association. 1971;**66**(334):243-250

[4] Eichhorn BH, Hayre LS. Scrambled randomized response methods for obtaining sensitive quantitative data. Journal of Statistical Planning and Inference. 1983;**7**(4):307-316

[5] Perri PF. Modified randomized devices for Simmons' model. Model Assisted Statistics and Applications. 2008;**3**(3):233-239

[6] Bhargava M, Singh R. A note on a modified randomization device using unrelated question. Metron-International Journal of Statistics. 1999; **57**(3-4):141-145

[7] Singh S, Horn S, Singh R, Mangat NS. On the use of modified randomization device for estimating the prevalence of a sensitive attribute. Statistics in Transition. 2003;**6**(4):515-522

[8] Batool F, Shabbir J, Hussain H. On the estimation of a sensitive quantitative mean using blank cards. Communications in Statistics-Theory and Methods. 2017;**46**(6):3070-3079

[9] Singh S. On the estimation of correlation coefficient using scrambled responses. In: Chaudhuri A, Christofides TC, Rao CR, editors. Data Gathering, Analysis and Protection of Privacy Through Randomized Response Techniques: Qualitative and Quantitative Human Traits, Handbook of Statistics-34. Elsevier; 2016

[10] Singh GN, Kumar A, Vishwakarma GK. Estimation of population mean of sensitive quantitative character using blank cards in randomized device. Communications in Statistics-Simulation and Computation. 2018. DOI: 10.1080/03610918.2018.1502779

[11] Singh GN, Kumar A, Vishwakarma GK. Some alternative additive randomized response models for estimation of population mean of quantitative sensitive variable in the presence of scramble variable. Communications in Statistics-Simulation and Computation. 2018. DOI: 10.1080/03610918.2018.1520879

[12] Narjis G, Shabbir J. An efficient partial randomized response model for estimating a rare sensitive attribute using Poisson distribution. Communications in Statistics-Theory and Methods. 2019. DOI: 10.1080/03610926.2019.1628992

[13] Narjis G, Shabbir J. Bayesian analysis of optional unrelated question randomized response models. Communications in Statistics-Theory and Methods. 2020. DOI: 10.1080/03610926.2020.1713367

[14] Ahmed S, Sedory SA, Singh S. Simultaneous estimation of means of two sensitive variables. Communications in Statistics-Theory and Methods. 2018;**47**(2):324-343

[15] Lanke J. On the degree of protection in randomized interviews. International Statistical Review. 1976;**44**:197-203

[16] Leysieffer RW, Warner SL. Respondent jeopardy and optimal designs in randomized response models. Journal of the American Statistical Association. 1976;**71**:649-656

[17] Bhargava M, Singh R. On the efficiency comparison of certain randomized response strategies. Metrika. 2002;**55**:191-197

[18] Diana G, Perri PF. Efficiency vs privacy protection in SRR methods. In: Proceedings of 44th Scientific Meeting of the Italian Statistical Society. 2008

[19] Zhimin H, Zaizai Y, Lidong W. Combination of the additive and multiplicative models at the estimation stage. In: 2010 International Conference on Computer and Communication Technologies in Agriculture Engineering. 2010. pp. 172-174

[20] Diana G, Perri PF. A class of estimators for quantitative sensitive data. Statistical Papers. 2011;**52**(3): 633-650

[21] Sarndal CE, Swensson B, Wretman J. Model Assisted Survey Sampling, Springer Series in Statistics. Springer-Verlag Publishing; 1992

Chapter 10

Causality Relationship between Import, Export and Exim Bank Loans: Turkish Economy

Yüksel Akay Ünvan and Ulviyya Nahmatli

Abstract

Export promotion tools aim to increase exports and support the entrepreneur in reaching new foreign markets. The positive impact of incentives, especially on financial issues, on exports both before and after shipment is undeniable. Founded in 1987, Turkish Exim bank is Turkey's official export credit institution. By observing macro-economic balances, Exim bank ensures that exporters, export-oriented production manufacturers and entrepreneurs operating abroad are supported by credit, guarantee and insurance programs to increase their competitiveness. The study aims to examine the causal relationship between imports, exports and Exim bank loans in the Turkish economy. In the study, stationarity with the extended Dickey-Fuller unit root test, long-term relationship with the Johansen co-integration test, and then causality with the Granger test were investigated. The causality relationship was analyzed using import, export and Eximbank loans data for the periods 2003–2020.

Keywords: exports, exim bank loans, ADF test, causality test

1. Introduction

For developing countries to reach the level of developed countries and to catch the level to compete with them, more than one condition must be met. The most important of these conditions is the industrialization strategies that developing countries will implement. With the decisions of January 24, 1980, which were a turning point in terms of redesigning the Turkish economy, the export-based industrialization strategy was started to be implemented by targeting export-based growth instead of the import substitution strategy implemented since the 1960s, and some institutions were created to eliminate the problems that will be encountered at the implementation stage of these decisions ([1], p. 22).

To increase the competitiveness of exporters in foreign markets, Turkish Exim bank provides export financing in Turkey with credit, guarantee and insurance programs under international rules and principles ([2], p. 180).

In developing countries, Exim bank loans are provided by organizations that support the Central Bank of the Republic of Turkey (CBRT) and non-profit exports. Commercial banks, private equity export credit insurance companies and factoring companies are the only organizations that support finance, as the main purpose is profit.

In developed countries, the necessary financing for exports is usually provided by the commercial banking system. Export financing organizations, on the other hand, support the export sector and banks with insurance and guarantee programs, only performs the function of providing a risk-free environment.

1.1 Import

Imports are the value of foreign goods and services bought by a country's households, firms, government agencies, and other organizations in a given period.

1.2 Exports

Exports are goods and services that are produced in one country and sold to buyers in another. Exports, along with imports, make up international trade.

1.3 Eximbank loan

Eximbank loans are lines of credit made available by Export Credit Bank of Turkey (Exim bank) to enhance exports. This credit is made available during the pre-export stage against a written pledge by the exporter to export Turkish-origin goods and services as stipulated by Exim bank. It provides a price advantage over other export loans offered by banks.

2. Literature review

In the Literature view, a summary of information was given about research that examines the relationship between exports, financial development and economic growth in Turkey in the context of causality.

Dodaro [3], examined the relationship between economic growth and exports with the Granger Causality test by using variables between 1967 and 1986 periods. The study found a one-sided causal relationship from economic growth to exports.

Bahmani and Domac [4] examined the relationship between economic growth and exports, with the Co-Integration test by using variables between 1923 and 1990 periods. As a result of the research, it is found that there is a decidedly causal relationship between economic growth and exports.

Tuncer [5], examined the causal relationship between exports, imports, investments and Gross domestic product (GDP) with the method Toda and Yamamoto by using variables between 1980Q1 and 2000Q3 periods. As a result of the study, a one-sided causality relationship has been found from economic growth to exports.

Şimşek [6], tested the export-based growth hypothesis with Error Correction Model, Co-Integration Test and Causality tests by using variables between 1960 and 2002 periods. As a result of the study, the one-sided causality relationship has been found from economic growth to exports.

Erdogan [7], examined the relationship between economic growth and exports, with Co-Integration and Causality tests by using variables between 1923 and 2004 periods. As a result of the study, the long-term double-sided causal relationship between economic growth and exports was found at the level of 10% significance.

Taştan [8], examined the interaction and causal relationships between export, industrial production and import variables, with Co-Integration and Causality tests by using variables between 1985Q1 and 2009Q3 periods. As a result of the study, a one-sided causality relationship has been found from economic growth to exports.

Tıraşoglu [9], examined whether the export-based growth hypothesis is valid in Turkey or not, with Co-Integration and Causality tests by using variables between

1998Q1–2011Q3 periods. As a result of the study, there is a long-term one-sided causal relationship between exports and economic growth.

Korkmaz [10], examined the relationship between economic growth and exports, with Co-Integration and Causality tests by using variables between 1998: Q1–2013:Q3 periods. As a result of the study, a one-sided causality relationship has been found from exports to economic growth.

Pentecost and Kar [11], examined the relationship between economic growth and exports, with Co-Integration and Causality tests by using variables between 1963 and 1995 periods. As a result of the research, there is a one-sided causal relationship from economic growth to financial development.

Al-Yousif [12], studied the causal relationship between financial development and economic growth for 30 developing countries, with both Time Series and Panel Data Analysis tests, by using variables between 1970 and 1999 periods. As a result of the study, there is a double-sided relationship between economic growth and financial development.

Ceylan and Durkaya [13], examined the causal relationship between domestic credit volume and economic growth, by taking advantage of Gross domestic product (GDP) and total loans that private banks use domestically by using variables between 1998 and 2008 periods. As a result of the research, there is a one-sided causality relationship from economic growth to loans.

3. Econometric analysis

3.1 Data set

In this study, the data set used were between 2003 and 2019 periods. The source of the data used in the study was taken from the Central Bank of the Republic of Turkey (TCMB) and the official website of the bank Exim bank. This data was created with three different variables which are listed in **Table 1**. All analyses and tests were performed on these variables by using the EViews11 program.

3.2 Augmented Dickey-Fuller (ADF) unit root test

To obtain econometrically significant relationships between series in time series analysis, it is essential that the analyzed series must be stationary. Unit root tests are usually used to test whether the series has a stationary structure or not. The most commonly used of these tests is the unit root test performed by Dickey-Fuller [14], which assumes that the error term is independent and uniformly distributed. If a time series is stationary, its variance, average, and covariance (with various delays) are the same, no matter when it is measured ([15], p. 757).

Variable	Code
Import	Central Bank of the Republic of Turkey, Balance Of Payments Analytical Presentation (6.manual), A.2
Export	Central Bank of the Republic of Turkey, Balance Of Payments Analytical Presentation (6.manual), A.1
Exim bank loans	https://www.eximbank.gov.tr

Table 1.
Data set.

Let Y_t be any time series, the stationary of a series depends on the following conditions:

$$\mathrm{E}\,(Y_t) = \mu \tag{1}$$

$$\mathrm{Var}\,(Y_t) = \mathbf{E}(Y_t\text{–}\mu)^2 = \sigma^2 \tag{2}$$

$$\gamma_k = \mathbf{E}[(Y_t\text{–}\mu)(Y_{t-k} - \mu)] \tag{3}$$

The relationship between this period value of Series Y_t and the value it has in the last period, is as in Eq. (4):

$$Y_t = \rho Y_{t-1} + \varepsilon_t \tag{4}$$

$$Y_t - Y_{t-1} = \rho Y_{t-1} - Y_{t-1} + \varepsilon_t \tag{5}$$

$$\Delta Y_t = (\rho - 1)Y_{t-1} + \varepsilon_t \tag{6}$$

$$\Delta Y_t = \gamma\ Y_{t-1} + \varepsilon_t \tag{7}$$

If $\rho = 1$ or $\gamma = 0$ is found in this equation, there is a unit root problem. If $\rho = 1$, the relationship will be as in Eq. (8):

$$Y_t = Y_{t-1} + \varepsilon_t \tag{8}$$

This means that the impact of the shock that the series was subjected in the previous period remains in the system as it was. If $\rho < 1$, it means that the initial effect of shocks in the past continues and that this effect will disappear over time.

The main regression patterns used in the Dickey-Fuller test are:

$$\Delta Y_t = \gamma\ Y_{t-1} + \varepsilon_t \tag{9}$$

$$\Delta Y_t = \beta_{0+}\gamma\ Y_{t-1} + \varepsilon_t \tag{10}$$

$$\Delta Y_t = \beta_{0+}\beta_i t + \gamma\ Y_{t-1} + \varepsilon_t \tag{11}$$

Eq. (9), shows a structure with no fixed term and no trend effect. Eq. (10) shows a structure with a fixed term and no trend term, and Eq. (11) shows a structure with a fixed term and no trend effect.

In case of correlation between error terms, the extended Dickey-Fuller (ADF) unit root test was developed again by Augmented Dickey-Fuller [16] by including the delayed values of the dependent variable in the model. The proposed models for this test are shown in the following equations:

$$\Delta Y_t = \gamma\ Y_{t-1} + \sum_{i=2}^{\rho} \beta_i \Delta Y_{t-i+1} + \varepsilon_t \tag{12}$$

$$\Delta Y_t = \beta_{0+}\gamma\ Y_{t-1} + \sum_{i=2}^{\rho} \beta_i \Delta Y_{t-i+1} + \varepsilon_t \tag{13}$$

$$\Delta Y_t = \beta_{0+}\beta t + \gamma\ Y_{t-1} + \sum_{i=2}^{\rho} \beta_i \Delta Y_{t-i+1} + \varepsilon_t \tag{14}$$

Eq. (12) shows the structure in which there is no fixed term and no trend effect. Eq. (13) shows the structure in which there is only a fixed term, and Eq. (14) shows the structure in which both the fixed term and the trend effect are observed.

The stationary test is first performed at the level value. If stationary is not achieved in the level value, the first difference of the Y_t series will be taken. If the

$\Delta Y_t = Y_t - Y_{t-1}$ series becomes stationary, it is denoted by I(1) and the series becomes stationary in the first difference. If stationarity cannot be achieved in the first difference of the series, the second difference will be taken. The process of taking the difference of the series continues until it becomes stationary.

In Eqs. (4) and (7), the H_0: $\gamma=0$ (the series aren't stationary) hypothesis in the unit root test was found by Dickey Fuller [14] and tested with the τ (tau) statistic. If the error term is correlated in the Y_t series, the extended Dickey Fuller (ADF) test is preferred, and the H_0 hypothesis is rejected if the critical values of MacKinnon [17], correspond to the absolute value of the statistics τ (tau), are greater than τ. ([15], p. 757).

If the ADF test statistic value is more negative than the MacKinnon [17] critical values at various significance levels, it is decided that there is a unit root in the series; in other words, the series are not stationary. In this study, the stability of the series was analyzed using the extended Dickey-Fuller (ADF) unit Root Test.

As we can see in **Table 2**, Import variables were found stationary in the intercept model in the first difference I(1), Export variables were found stationary in non-intercept and trendless model in the first difference I(1); while Eximbank loans variables were found stationary in intercept model in the second difference I(2).

3.3 Johansen cointegration test

To test whether non-stationary series converge to equilibrium over a long period, the cointegration test examines whether there is a long-term relationship between the series or not. But since this test does not provide information about the direction of the relationship, causality tests are used to determine the direction of the relationship. There are two Tests in Johansen's cointegration analysis. These are trace and max.

Trace hypothesis test H0: $r \le r0$, H1: $r \ge r0 + 1$.

Max hypothesis test H0: $r = r0$, H1: $r = r0 + 1$.

If $r = 0$ there is not cointegration vector.

The series were analyzed using the Johansen cointegration test and the results were shown in **Table 3**. In **Table 3**, the $r = 0$ hypothesis, shows that there is no cointegration relationship between the variables; the $r \ge 1$ hypothesis, is an alternative hypothesis which shows that there is at least one cointegration relationship; the

Variables	Test for unit root in	Include in test equation	Lag Length	ADF
Import	I(1)	Intercept	p = 0	−4.061237
Export	I(1)	None	p = 0	−3.196258
Eximbank loans	I(2)	Intercept	p = 1	−4.417361

Table 2.
ADF unit root test.

Hypothesis		Trace statistic		Max-Eigen statistic	
H_0	H_1	Statistic	Critical value	Statistic	Critical value
r = 0	r ≥ 0	76.02502	29.79707	66.68893	21.13162
r = 1	r ≥ 2	9.336092	15.49471	8.804522	14.26460

Table 3.
Johansen cointegration test results.

r ≥ 2 hypothesis is an alternative hypothesis that shows that there are at least two cointegration relations:

According to the Johansen test output, both the Trace test statistic value and the Maximum Eigen test statistic value were greater than the table critical value of 5%. Therefore, the zero hypothesis of r = 0 can be rejected for both test values. In other words, Export, Gross domestic product (GDP), and Loan variables are cointegrated.

3.4 Granger causality test

The Granger causality test examines the relationship between series based on estimating past and present values. According to Granger, if past information about X_t helps to obtain estimates. On the other hand, if Y_t's past values allow X_t to be estimated, the Y_t series is the granger cause of X_t. If X_t causes Y_t and Y_t causes X_t, there is a bilateral causality relationship. An error correction model is used to determine the direction of the causality relationship, if the series is co-integrated. But if the series is not co-integrated, standard Granger or Sims tests are used to determine the direction of the causality relationship ([18], pp. 213–228).

3.4.1 Determination of appropriate lag length

Accurate determination of the number of lag lengths in the Granger causality test is very important for the application to give healthy results, because this test is sensitive to the number of lag lengths. To find the appropriate lag length numbers for the Granger causality test, the Vector autoregression (VAR) model is estimated. Here a generic VAR model is estimated primarily to determine the appropriate number of lag length. Then, the number of lag length, will be determined by Akaike information criteria and by the LM test.

For the VAR model, the appropriate lag length was obtained by LogL (Log-We), LR (sequential modified LR test statistic), FPE (Final prediction error), AIC (Akaike information criterion), SC (Schwarz information criterion) and HQ (Hannan-Quinn information criterion) criteria. The model with the largest LogL and LR values and the smallest FPE, AIC, SC and HQ values were selected to determine the appropriate lag length criteria.

As seen from **Table 4**, Sequentially modified LR test statistic (LR); Final prediction error (FPE), Akaike information criterion (AIC),Schwarz information criterion (SC) and Hannan-Quinn information criterion (HQ) appropriate lag length as 1. According to this information, the lag length will be 1.

In **Figure 1** it is presented the Var(1) model which provides the stationary condition:

Since the auto-regressive characteristic roots are all in the unit circle,the model VAR(1) which is used in the study, provided the stationary condition. Subsequently, appropriate delay numbers for the Granger causality test were performed

Lag	LogL	LR	FPE	AİC	SC	HQ
0	−853.9328	0	6.65e+42	107.1166	107.2615	107.1240
1	−820.4059	50.29035*	3.20e+41*	104.0507*	104.6302*	104.0804*
2	−816.0730	4.874522	6.71e+41	104.6341	105.6481	104.6861

**Values shows that the appropriate number of lag lengths according to the relevant criterion.*

Table 4.
Determination of appropriate lag length.

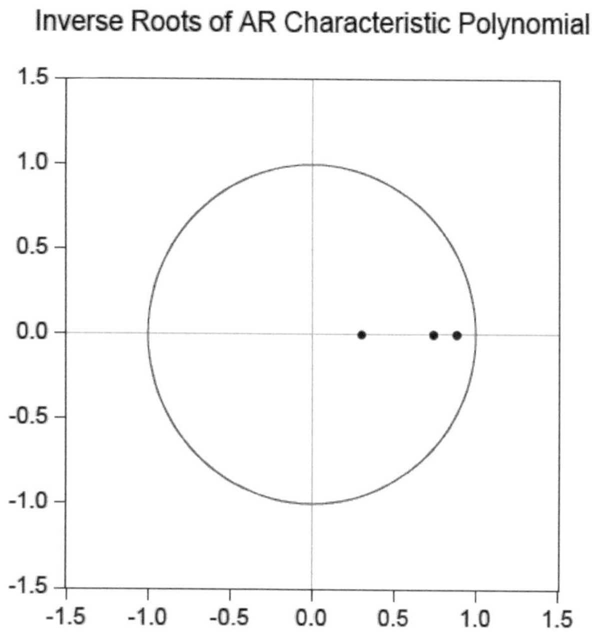

Figure 1.
Stationarity analysis.

H_0 Hypothesis	Chi-sq.	Probability	Result
Eximbank→Export	1.331500	0.2485	Rejected
Import→Export	2.497195	0.1140	Rejected
Export→Exim bank	0.762369	0.3826	Rejected
Import→Eximbank	2.986957	0.0839	Rejected
Eximbank→Import	0.000609	0.9803	Rejected
Export→Import	0.017613	0.8944	Rejected

Table 5.
Granger causality test results.

by autocorrelation LM tests, it was determined that there was no autocorrelation and the series was stationary.

The series were analyzed using the Granger causality test, as we can see from **Table 5**; there is no causal relationship between Eximbank to Export variables (ρ = 0.2485 > 0.05), Import to Export variables (ρ =0.1140 > 0.05), Export and Eximbank variables(ρ = 0.3826 > 0.05), Import to Eximbank variables (ρ = 0.0839 > 0.05), Eximbank to Import(ρ =0.98035 > 0.05), Export to Import (ρ =0.8944 > 05).

According to the results which are shown in **Table 5**, it was determined that there is no causal relationship between Eximbank loans, Import and Export variables at 1 and 5% significance levels.

4. Conclusion

To decipher the causal relationship between import, export and Eximbank loan variables in the Turkish economy, three different variables were used in the study.

All variables used in the study are time series, because they depend on time, so the stationarity of the variables was tested by the ADF test. As a result of the test, stationarity was achieved by taking first-order differences in import and export variables and second-order differences in eximbank loans variables. To test whether non-stationary series converge to equilibrium over a long period or not, the series were analyzed by using the Johansen cointegration test and the results revealed that Export, GDP, and Loan variables were cointegrated. Then the series were analyzed using the Granger causality test, and according to the results, it was determined that there was no causal relationship between Eximbank loans, Import and Export variables at 1 and 5% significance levels.

When we look at the literature review, a summary of information was given about research that examines the relationship between exports, financial development and economic growth in Turkey in the context of causality. From the study of Ceylan and Durkaya [13], there was found one-sided causality relationship from economic growth to loans. From the study of Dodaro [3], Bahmani and Domac [4], Tuncer [5], Şimşek [6] and Taştan [8] it was found a causal relationship from economic growth to exports. Erdogan [7] found causality relationship between economic growth and exports at the level of 10% significance. Tıraşoğlu [9] and Korkmaz [10], found a causal relationship between export and economic growth. Pentecost, Kar [11] and Al-Yousif [12] found causal relationships from economic growth to financial development. But in this study, it was determined that there were no causal relationship between Eximbank loans, Import and Export variables at 1 and 5% significance levels.

Turkey's export target in 2023, is to set at 500 billion USD. Looking at the export figures at the end of 2015, Turkey must increase exports by an average of 16.5% each year to reach the 2023 target. To achieve this increase, it is necessary to ensure the high growth of the economy, accelerate R&D investments, diversify exports, reach new markets, and provide the necessary regulations and facilities for exporting companies to compete with exporters in other countries.

Eximbank loans provide a price advantage over other export loans offered by banks. It has a strong financial structure. Because of this financial structure, it supports exports at a high rate. To achieve the export potential that the country has, also in international markets, it should implement new and effective credit/insurance programs under international treaties and the restrictions of the institutions to which it is affiliated.

Author details

Yüksel Akay Ünvan* and Ulviyya Nahmatli
Ankara Yıldırım Beyazıt University, Turkey, Ankara

*Address all correspondence to: aunvan@ybu.edu.tr